Aesthetics of Participation

This book explores the way people participate with the Oslo Opera House, Norway. As an iconic and culture-led building, these different modes of participation reveal the tensions between staged space and individual experience.

Movement, materiality, light, and art are viewed through an atmospheric lens to demonstrate how architecture can shape people's engagement with, and understanding of, urban space. This book contributes to a growing literature on atmosphere in relation to our experience of the built environment. In adopting this atmospheric perspective, the book speaks to the concerns of designers, users, and researchers interested in the way contemporary development infuses our cities with the experiential, as a means of developing access, participation, and democracy. It explores the ways in which people experience a building, held up against the claims, intentions, and assumptions that surround it.

The book's focus on design, participation, and experience, in relation to political ideals, will appeal to architects, planners, and academics concerned with the production of space. Equally, its underlying atmospheric contribution and methodological approach will be of interest to designers, scholars, professionals, and students of ambiance, affect and atmosphere, architecture, city planners and urban developers, human geographies, anthropology, and urban studies.

Jeremy Hektor Payne-Frank holds a PhD from the Department of People and Technology at Roskilde University. His research explores urban experience in relation to architecture, art, and design through experimental ethnographic methods.

Ambiances, Atmospheres and Sensory Experiences of Spaces

Series Editors:

Rainer Kazig, *CNRS Research Laboratory Ambiances – Architectures – Urbanités, Grenoble, France*
Damien Masson, *Université de Cergy-Pontoise, France*
Paul Simpson, *Plymouth University, UK*

Research on ambiances and atmospheres has grown significantly in recent years in a range of disciplines, including Francophone architecture and urban studies, German research related to philosophy and aesthetics, and a growing range of Anglophone research on affective atmospheres within human geography and sociology.

This series offers a forum for research that engages with questions around ambiances and atmospheres in exploring their significances in understanding social life. Each book in the series advances some combination of theoretical understandings, practical knowledges and methodological approaches. More specifically, a range of key questions which contributions to the series seek to address includes:

- In what ways do ambiances and atmospheres play a part in the unfolding of social life in a variety of settings?
- What kinds of ethical, aesthetic, and political possibilities might be opened up and cultivated through a focus on atmospheres/ambiances?
- How do actors such as planners, architects, managers, commercial interests and public authorities actively engage with ambiances and atmospheres or seek to shape them? How might these ambiances and atmospheres be reshaped towards critical ends?
- What original forms of representations can be found today to (re)present the sensory, the atmospheric, the experiential? What sort of writing, modes of expression, or vocabulary is required? What research methodologies and practices might we employ in engaging with ambiances and atmospheres?

Aesthetics of Participation
Atmosphere, Design, and Experience at the Oslo Opera House
Jeremy Hektor Payne-Frank

For more information about this series, please visit: www.routledge.com/Ambiances-Atmospheres-and-Sensory-Experiences-of-Spaces/book-series/AMB

Aesthetics of Participation

Atmosphere, Design, and Experience at the Oslo Opera House

Jeremy Hektor Payne-Frank

Routledge
Taylor & Francis Group
LONDON AND NEW YORK

First published 2025
by Routledge
4 Park Square, Milton Park, Abingdon, Oxon OX14 4RN

and by Routledge
605 Third Avenue, New York, NY 10158

Routledge is an imprint of the Taylor & Francis Group, an informa business

British Library Cataloguing-in-Publication Data
A catalogue record for this book is available from the British Library

Library of Congress Cataloging-in-Publication Data
Names: Payne-Frank, Jeremy Hektor, author.
Title: Aesthetics of participation / Jeremy Hektor Payne-Frank.
Description: Abingdon, Oxon ; New York, NY : Routledge, 2025. | Series: Ambiances, atmospheres and the sensory experiences of space | Revision of the author's thesis (doctoral)--Roskilde University, 2022. | Includes bibliographical references and index.
Identifiers: LCCN 2024042181 (print) | LCCN 2024042182 (ebook) | ISBN 9781032668772 (hardback) | ISBN 9781032668789 (paperback) | ISBN 9781032668802 (ebook)
Subjects: LCSH: Operahuset (Oslo, Norway) | Snøhetta (Firm) | Architecture--Human factors--Norway--Oslo. | Space (Architecture)--Norway--Oslo. | Oslo (Norway)--Buildings, structures, etc.
Classification: LCC NA6840.N82 O856 2025 (print) | LCC NA6840.N82 (ebook) | DDC 725/.822097821--dc23/eng/20241019
LC record available at https://lccn.loc.gov/2024042181
LC ebook record available at https://lccn.loc.gov/2024042182

ISBN: 978-1-032-66877-2 (hbk)
ISBN: 978-1-032-66878-9 (pbk)
ISBN: 978-1-032-66880-2 (ebk)

DOI: 10.4324/9781032668802

Typeset in Times New Roman
by SPi Technologies India Pvt Ltd (Straive)

For Selma, Aslan, and Eila.

Contents

List of Figures *ix*
Preface *xi*

1 Introduction: The Oslo Opera House 1
A social monument 4
Tourists and locals 7
An atmospheric lens 7
The Nordic invitation to participate 10
Immaterial architecture 11

2 Adopting an atmospheric lens 16
Atmospheric perspectives: Böhme, Rancière, and Sloterdijk 19
Thinking about atmospheres 20
Building experience: architecture and atmosphere 23
Coercive atmosphere 26
Research atmosphere: methods and tools 29
Withdrawing from atmosphere: taking photography seriously 32
Pinholes and fuzzy 34

3 Transformative participation 39
The road to Bjørvika 42
The Fjord City 44
Bjørvika and beyond 46
Snøhetta 48
A Nolli map 51
Competition entry 04321 53
The art 54
A stone saga 56
Rethinking participation: Arnstein's ladder 57

4 Material participation 61
From the city to the roof *61*
Marble: surfaces of the white carpet *65*
Whiteness *67*
Marble's social and synaesthetic character *69*
Whiteness and a cleansing of the eye *77*
Whiteness and blur *78*
Designing ambiguity: the palace, whiteness, and a Norwegian sensibility to nature *80*

5 Movement participation 83
Architecture of the oblique *87*
Dwelling, tripping, inhabiting *92*
Life in the Norwegian open air *97*
Resonance, dissonance, and a good-natured elitism *102*

6 Light participation 105
Nordic light, Nordic architecture: from the roof to the foyer *106*
Daylight in the OOH *110*
Artificial light *114*
Transitions: liquid light *118*
From bubbles to foam *123*

7 Art participation 126
First encounter with The other wall *128*
Democratic surfaces *129*
Democratic surrounds: The wall *as weather machine* *134*
Creative kitchens: process as art *135*
The wall *through social media* *137*
Selfies and mirror selfies *144*
The wall, *dissensus, and a partitioning of the sensible* *145*

8 Conclusion: Exiting the social monument 148

Bibliography *156*
Index *170*

Figures

1.1 The OOH attracts a variety of visitors and activities. Photograph by the author. 2
1.2 A sunset attracts visitors to the social monument. Photograph by the author. 4
1.3 Visitors enjoying the view of the fjord from the roof of the OOH. Photograph by the author. 6
1.4 Children playing in the foyer. Photograph by the author. 10
2.1 Thinking ontologically: A pencil sketch of co-mingling spheres. By the author. 17
2.2 Learning to look for the elements of atmospheric. A fieldwork sketch of the OOH's roof. By the author. 31
2.3 A pinhole image showing people moving around the OOH. Photograph by the author. 35
2.4 A pinhole image showing natural light interaction with the foyer space. Photograph by the author. 36
3.1 Construction work around the OOH with the *Deichman* library behind. Photograph by the author. 40
3.2 An architectural vision of future Bjørvika covers the site of its construction. Photograph by the author. 45
3.3 An architectural model of the OOH displayed inside Snøhetta's studio. Photograph by the author. 50
3.4 Costume designers display their craft to passers-by. Photograph by the author. 55
4.1 Viewed from above the scale and complexity of the 'white carpet' is revealed. Photograph by the author. 62
4.2 Hammered, grooved, and cut marble. Photograph by the author. 64
4.3 Some stains are quickly removed. Debris on the 'beach' after a winter gale. Photograph by the author. 66
4.4 The marble produces snowlike qualities in the low Nordic light. Photograph by the author. 75
4.5 A painting by Harald Sohlberg entitled, 'Winter Night in the Mountains' (1914). Public domain. 76

4.6 Wet marble creates an ice-like landscape. Photograph by the author. 79
5.1 Composite of a time-lapse shows the OOH as it transitions into darkness. Photograph by the author. 85
5.2 A rendering of the OOH's logo. Photograph by the author. 88
5.3 Bodily participation at the OOH. Photograph by the author. 90
5.4 People making their own journey: visitors migrate towards the sunset at the top of the OOH's roof. Photograph by the author. 97
5.5 Scooter riders appropriate the roof as a utility space. Photograph by the author. 100
6.1 A composite of photos taken inside the OOH's foyer. Photograph by the author. 106
6.2 *The other wall* and the oak-clad *wave wall*. Photograph by the author. 108
6.3 Silhouettes of activity on the roof, cast upon the wooden *wave wall*. Photograph by the author. 112
6.4 At night the OOH uses light to become a lantern. Photograph by the author. 114
6.5 The cold tone of *The other wall* is contrasted by the warmth of the wood wall. Photograph by the author. 116
6.6 The sunset designed by ÅF. An example of Oslo's many urban light designs. Photograph by the author. 120
6.7 *Akrobaten* pedestrian bridge with the *Barcode* development behind. Photograph by Bjørn Joachimsen. 122
7.1 Detail of *The wall's* geometric structure. Photograph by the author. 127
7.2 In darkness *The wall's* light begins to interact with the foyer's materials and form. Photograph by the author. 128
7.3 Two young visitors framed by shadow lean against *The other wall* and have their picture taken. Photograph by the author. 133
7.4 *The other wall* altered through photography to a saturated and sculptural form. Photograph by Bjørn Joachimsen. 141
7.5 The silhouette of a figure with a cap and trench coat is mirrored against the vibrant colours and geometrical structure of *The other wall*. Photograph by Morten Andresen. 143

Preface

This book is the result of my part in the research project, 'Living with Nordic Lighting', funded by the Velux foundation from 2018 until 2022 at Roskilde University. At that time, my designated area of study was architecture and light, with the expectation that I would find cases in Denmark, Sweden, and Norway. However, with the onset of the COVID-19 pandemic, three became two, became one, and I was left to focus my energy on the Oslo Opera House. As it turned out, its iconic, culture-driven architecture was ample material for study and its waterfront location, array of actors, and diverse functionalities threw forth themes and questions about how, why, and for whom Oslo was being developed.

As a Nordic planning imperative, a 'people first' approach to architecture and development attracts international attention. Related themes such as democratic public space, access, and citizen participation can be applied to the Opera House, yet my initial impression of the Oslo Opera House was that it was still something of an outlier. As I crossed the small bridge from the city and entered its expansive outdoor area, I found a unique space that summoned the senses and invoked a corporeal experience seemingly shared by the other bodies present. Nevertheless, I was left wondering what it was that people were doing in this space as they wandered around *exploring*, what attracted them, and how were they engaged with/by the space.

Looking at the way architecture might *stage* the way people feel, think, and do, led me to explore participation as played out in a building both designed and felt. At an early stage of my research, I came upon the claim that the Opera was a kind of 'social monument'. This paradoxical combination of positions intrigued me, eventually underpinning my inquiry, and I began to explore the tensions between our ability to appropriate space, and the forces that might guide us along certain *paths*. Thus, participation, and the invitation to participate, became the locus of my interest as it was here agendas, intentions, and experiences would meet to reveal something about people's relationship with contemporary architecture, design, and urban development.

Despite the disruptions of the Corona pandemic, I was able to conduct five weeks of ethnographic fieldwork. I visited Oslo on five occasions, spanning different seasons to capture the varying conditions and feelings of place.

Drawing from atmospheric theory my research began by uncovering the expressions of visitors and their situated perspectives. As a British citizen living in Denmark, I found myself straddling the local and the global, arriving with insights into Nordic sensibilities, as well as an awareness of tourist perceptions. As time went on, experiences in the field led me to consider the notion of Norwegianness, particularly in the way it was being constituted through emotional responses to the OOH's assemblage of material and immaterial elements that were grounded in cultural understandings. In this sense, my contribution to the notion of the Nordic became an interrogation into the Opera's staging of ambiguity, exemplified by Nordic light and its deep cultural associations, related practices, and brandable aesthetics.

Such a task is necessarily cross-disciplinary, and I was drawn to an eclectic mix of literature from art, architecture, and social sciences. That said, the book retains an atmospheric perspective that compels reflection around the role of research in a space that invites us to participate. Consequently, the notion of atmosphere remains elusive, and I attempt to embrace ambiguity as part of an analysis of built and experiential space.

Throughout this journey, I hope to have been a valuable and amiable colleague, productively contributing to the atmosphere of research. And, in this context, I would like to acknowledge the invaluable contribution Mikkel Bille, David Pinder, Siri Schwabe, and Annette Stenslund have made to the work that culminated in this book. Thanks, should also go to Sarah Pink, Tim Edensor, Tim Flohr Sørensen, and Katja Bülow for the useful exchange of ideas and feedback, as well as Mick Frank, Jonas Larsen, and Jørgen Ole Bærenholdt for their support and advice. Lastly, I'm indebted to Monica Degen and Andreas Philippopoulos-Mihalopoulos for encouraging me to transform my research into the volume you now have before you.

1 Introduction

The Oslo Opera House

My first encounter with the Oslo Opera House (OOH) was in the spring of 2019. Oslo was alluring because it was in a process of reclaiming large parts of its harbour front, transforming disused industrial areas into commercial, cultural, and domestic extensions of the city. I first visited Aker Brygge and Tjuvholmen, two completed zones of Oslo's extensive 10-km-long *Fjordbyen* development[1] (Vibe 2007). Architecturally, they feature the Astrup Fearnley Museum of Modern Art from 2012, a delicate wood clad building designed by Italian architect Renzo Piano that straddles the water like an odd sailboat. There was a public café on one side and a small beach further up the dock where a few brave locals were taking a spring dip. The activity provided something out of the ordinary for a foreign visitor who felt pleasantly at ease but was otherwise unmoved by the strangely familiar scene. Architect Cedric Price once wrote that we should not worry about places looking the same, only that they look so bloody (Hughes & Sadler 2000, 24). Price was writing about the under-invested roughness of English towns in the 1970s, whereas here I was presented with a clean and functional aesthetic reminiscent of many other contemporary harbour front developments.

Moving back to the centre of the town I arrived at Bjørvika, another part of Oslo's Fjord City under heavy development at the time. Flanked by construction work I encountered the OOH, a building that *felt* altogether different from what I had experienced earlier that day. Its scale alone provided an expanse of space that I found physically awakening but not unsettling. Its surface was covered in a white stone that was clearly worked with intention, creating a *carpet* that covered its entirety. Differently sized rectangles interlocked to form a minimalistic platform that extended upwards in engaging geometries. What first appeared as quite simple, clearly was not. There were people, many people, all moving in a way that seemed explorative, operating at a different tempo than in the city. The space was well used, yet people were seemingly doing nothing in particular. Children seemed compelled to run into the openness, and laughter could be heard from around. The mood was light, the roof surroundings apparently triggering in others, as it had for me, a kind of response. What might these casual, seemingly insignificant interactions with a building amount to? Despite the architecture's recognisable global style, the

DOI: 10.4324/9781032668802-1

site felt Nordic: an understated lack of ornament, open and transparent, its use of wood and stone materials, the allusion to nature. These were some of the scant explanations I could offer from my first encounter; nevertheless, they amounted to a kind of feeling I could not deny. The place exuded an atmosphere that was both familiar and yet novel, everyday yet unique, modest yet luxurious. These initial impressions formed my underlying curiosity and motivation for this book: a general wonder about what it is that people are doing and feeling when they are invited to participate with these types of staged spaces.

> There is a new house in the city. Everyone is talking about it. Now, when you come to Oslo, you have to see this house. And not just see it, even better, walk on and over it, around and in it. Like an anthill, I thought in those first days when they opened the building to outsiders, and it sucked in both citizens and guests.
>
> (Olin 2008)

The quotation above was written around the time the OOH opened to the public in 2008. It highlights the importance of the building to its local population and its role as a city icon; moreover, it signals the architecture's experiential qualities, 'sucking' people onto its architectural stage and encouraging them to walk *on* it, not merely look at it. It's a building that demands our participation.

Figure 1.1 The OOH attracts a variety of visitors and activities. Photograph by the author.

The quote also speaks to a building bestowed with multiple roles, since defined as: a city brander, tourist developer, mobility enhancer, national identity builder, and nurturer of local economies (Smith & Strand 2011, 94). Snøhetta, the local architects behind the OOH's design, deem their work to be a type of 'social monument' (Dykers 2012), and 'not an icon' (Thorsen & Obrist 2011, 56), inferring an open, democratic, and public space that combines the two basic spatial compositions in architecture: 'the closed architectural body, which delimits, and the open body, which embraces an area that is connected with infinite space' (Zumthor quoted in Böhme 2016, 138).

Another way of expressing this overlapping dichotomy, one often used in anthropology and human geography, is the relationship between *place* and *space*. As with Zumthor's understanding of architectural types, place tends to refer to bounded areas of space and time, whereas space, taken alone, is an unbounded continuum (Malpas 2021, 5). It is difficult to think of architecture and urban development in any way other than being typologies of place because of the human initiatives that alter the particular environment. Yet, as we will see in the following chapters, the notion of space as an open continuum remains present, summoned through designed interventions that make the boundaries of place more, or less, porous. It is pertinent then to think about how architectures make us *feel* connected to other spatialities, especially when it is achieved through artifice, and how certain meanings arrive with these other places, summoned across multiple temporalities. When the design of a place touches an individual, I refer to *atmosphere* as the way staged environments are decoded through the subjects' memories, cultural understandings, and expectations, 'in a border zone between sensuousness and meaning' (Albertsen 2019, 4).

Thinking atmospherically highlights a resonance with something exterior to the structure we are in, and how we might relate to, or *dwell* with, the built. Yet, atmosphere is not only composed of resonances; dissonance can also be shaped, guiding us, or warning us, encouraging us to feel in some way about something. So, dissonance is equally related to atmosphere, but in architecture where 'the building really is the world' (Jameson in Leach 1997, 253), dissonance tends to be designed out of experience in the pursuit of ideal human conditions.

One aspect of what makes the OOH a compelling case for study is that its 'generous' open public spaces suggest a loosening of architectural control and with it the potential for diverse atmospheres to emerge. Nevertheless, the notion of atmosphere within design circles is usually tacit rather than stated. For example, in urban development, what is often referred to as 'amenity value' is exactly the notion of what a place will feel like to its future users. Explicit examples can also be found[2], and the degree to which designers are able to coerce users in subtle ways shows atmosphere to be fundamental to revealing the power dynamics within contemporary urban development.

This book follows these lines of thought, investigating how the material and immaterial arrangement of a building reifies our relations with a space. How our participation with place is not only the act of participation but also a feeling of connection, an *attunement* with the world that comes via the staging of space. This position does not necessarily denote design as deterministic, but an attempt to think with and about the constitution of atmosphere, how it *feels*, and generates certain understandings in response to an environment that is layered and complex. In so doing, this book aims to contribute to an expanding body of literature concerned with atmosphere as a theoretical concept, a methodological tool, and as a means for better understanding the role atmospheres have in shaping aspects of our everyday lives.

A social monument

In the first three years of its opening, the OOH had attracted an estimated five million visitors exceeding all expectations (Berglund 2020). Its sheer popularity supports claims of an 'egalitarian, democratic and non-commercial use' of space (Fløgstad 2010, 62), furthering the general opinion that the building's architects Snøhetta have successfully designed an iconic building, 'loaded with metaphor' yet maintaining an 'aura of being 100 percent public' (Nielsen 2012, 181). However, the multiple roles and claims surrounding the OOH also raise questions about the nature of its democratic public space, where the contrast between iconicity and an architecture concerned with people holds a tension between the passively perceiving and actively experiencing individual. The 'social monument' frames two seemingly oppositional positions: between participatory emancipation and what Rancière calls the 'stultification' of the spectator (2011).

Figure 1.2 A sunset attracts visitors to the social monument. Photograph by the author.

It is often assumed that to participate is to obtain a degree of agency which is in turn democratising. Its nemesis is the passive gaze, a paralysing encounter that cajoles the individual into giving up their *voice* and consume whatever is placed before them. Thus, the 'social monument' brings together oppositional terms, loaded with tensions between architecture's democratic potential and its relationship with capital and authority. Moreover, it foregrounds how participation is dependent on the arrangement of material elements that are themselves permeated by relations of power and constructions of citizenship (Cornwall 2002). Therefore, we should not assume the OOH to be a space wherein people enter free from compulsion, and it should be highlighted that 'the politics of constructed spaces does not end with the erection of buildings. This is, one might say, only its beginning' (Rockhill 2016, 258).

Consequently, when I use the term aesthetics, I am not referring to an aesthetics of taste but of experience; an *aisthesis* that infers the conditions that enable the object to be felt through form, rhythms, pauses, and glimmering surfaces (Rancière 2019a, x). Relatedly, I posit participation as different from experience as it also holds the promise of agency for the individual. In other words, the participant is a *user* of a place charged with intent, and I employ participation as a means of exploring the OOH's spatial production amidst the ambiguities of architectural design and individual agency. In this regard, I consider the materiality that flows through the space as much as the way situated bodies move (Degen & Rose 2022, 5) around the OOH's architecture and adopt an atmospheric perspective that holds spatial experience to be a co-mingling between object and subject (Böhme 1993, 114). I, therefore, attempt to give equal attention to the building's design, and the experience of that design, to better understand the ways we participate with our built environments.

In a general sense then, this book explores the ways in which people experience a building, held up against the claims, intentions, and assumptions that surround it. Keeping the ambiguous declaration that the OOH is 'a social monument' in mind, I unpack the tensions between the OOH's iconicity and its experiential qualities. If contemporary architecture tends to be judged by its ability to engage people with regard to what it makes you feel not what it is, then the way people are encouraged to participate in time and space (see Dezeuze, 2010, 4) is surely a valuable area for research. And we should ask, if an architecture is being lauded as a social monument, what kind of social is in fact being played out? In order to approach this question, I explore the *modes* of people's participation. I use the term modes intentionally, to hold open the possibility of different forms and varieties of types to be included in the analysis of participatory experience. Furthermore, modes relate to an interest in the urban environment as multimodal, involving various layers of mediating technologies that contribute to, but also challenge, conceptions of situated atmospheric experience.

Guided by these issues, I endeavour to analyse people's engagement with the architecture of the OOH without overlooking the potential for participation to exceed the building's spatial structure. I locate certain *modes of participation* that take place at the OOH and proceed by embracing the interdisciplinary nature of my case, drawing upon a wide range of influences that include architecture, art, human geography, and urban studies concerned with spatialised participatory practices. In so doing, I hope to contribute to understandings of the relationship between architecture and our urban experiences and the relatively understudied area of how individuals in fact participate with public space (Stevens & Lossau 2015), especially when it involves atmosphere's relation to the mediations of art, architecture, and social media. In addition, this book speaks to wider debates concerned with power dynamics in architecture and urban development, and the tensions between individual agency and atmospherically staged democratic space. And, in this regard, I point to the validity of including different perspectives when looking at spaces that are themselves the outcome of interdisciplinary practices. As a researcher, being *in*, talking *about*, and thinking *through* atmosphere (Sumartojo & Pink 2018a) lead to a similarly interdisciplinary position that is often transformative in the way it can produce new forms of knowledge (Moran 2006, 16). Accordingly, adopting an atmospheric perspective requires seeing participation through the entities from which they are produced (Böhme 2017, 79). Thus, atmosphere allows us to look empirically at the various configurations between design and experience and can potentially reveal where changes in the design process can be made (Sumartojo & Pink 2018b, 127).

Figure 1.3 Visitors enjoying the view of the fjord from the roof of the OOH. Photograph by the author.

Tourists and locals

To define the people targeted in my study, I make a distinction between the 'sequential' and 'synchronic' functionality of a space (Sennett 2018, 206). Due to my concern with participation and public space, I ignore the 'sequential' organisation of visitors arriving with a ticket, focusing instead on those who must navigate the OOH's 'synchronic' space. In other words, this book focuses on the open and freely accessible areas of the OOH's public space, namely, the building's foyer and roof area. Furthermore, the unique event style nature of visiting the OOH, what the architects Snøhetta refer to as 'Sunday architecture', means that the difference between local and foreign visitors becomes problematised as locals become tourists in their own city. In John Urry's seminal writings on the 'tourist gaze', he posits tourism as the difference between the ordinary and the extraordinary (Urry & Larsen 2011, 15). Yet, the OOH's multiple roles disrupt this easy distinction. Oslovians have been incorporating the building into their psyche for some years now, passing it on the way to work or on a leisure run through its surrounding district of Bjørvika. It, therefore, occupies a strange space between a tourist site and the everyday; an event space familiar to many citizens, who often refer to the building affectionately as a 'house'. For this reason, I will not define tourists and nontourists as different categories but instead concentrate on people's situated understandings and perceptions that also include memories, pre-conceptions, and even local identity. Here, I am also referring to the way foreign visitors bring notions of Nordicness and Nordic culture that are taken for granted by locals. In my conception, this universalised perspective of the Nordic replaces the 'tourist gaze' as the site of spatial production. It shapes 'an imagined world at which the natives grasp themselves as grasped from the outside' (Sloterdijk 2014, 30). I take this conflation of local and global as the common starting point for all visitors, against which individuals can potentially shape their own meanings and interpretations. Moreover, as Rem Koolhaas (2002, 181) has written, contemporary architecture brings new forms of consumption: 'half the population produces new space; the more affluent half consumes old space'. And whilst the elitist nature of the OOH's art and culture 'old space' is inevitably part of the story, my focus is on the atmospheric unfolding of its 'new space', its modes of participation, and how it is able to assuage multiple uses, intentions, and claims.

An atmospheric lens

In March 2020 the Corona pandemic reached Europe, bringing with it travel restrictions, lockdowns, and social distancing that meant a significant barrier to accessing the OOH. The impact of Corona led me to utilise the social media platform Instagram as a powerful insight into people's aesthetic sensibility. Seeing how people's interactions with the OOH are multimodal, relates to an atmospheric perspective. It enabled me to consider the way technology has changed the way we interact with our urban environments, and the way participating with architecture includes mediation via mobile devices (see Degen and Rose 2022, 35). This is further evidenced by the way institutions have embraced social media to the extent that we now talk of spaces as, 'Insta-worthy', 'Instagram-friendly', 'Instagram-ready' (MacDowall & Budge 2022, 58). The instagramability of a place blends the experiential with an invitation to participate by taking *the* picture. Therefore, social media produces a mediation of space by constituting how people might imagine a space even before their arrival and engages them in embodied ways (Degen & Rose 2022, 10) as they move around the OOH.

I combined social media data with in-situ interviews, talking to visitors, staff, and some of the experts who were involved in the process of designing the OOH, notably architects and artists involved in the design of the building. During fieldwork, my ethnographic approach included photography, sketching, video recordings, time-lapse photography, and participant observation, over five separate weeks of on-site fieldwork. The new Deichman library in Oslo, part of the Bjørvika development and neighbour building to the OOH, proved to be a fertile resource for newspaper articles and local planning documents, whilst secondary materials relating to Oslo's greater waterfront development are in the public domain and available online.

A further point to make about an atmospheric perspective relates to the researcher's own positionality. Understanding that data is the representation of the world, not the world itself (Dicks et al. 2006, 78), means the way research constitutes atmosphere should be taken into consideration. This reflection was a vital part of my methodological framework that developed over time as I considered, and felt, my own participation in the shaping of what I was examining, being a part of what Sarah Pink calls 'ethnographic place' (2015, 48). This was particularly apparent in my use of photography where taking a *good* image of the OOH was to submit to the dominant aesthetics of architectural photography, bound up with architectural images of dissemination, and including a tacit understanding of 'knowing where to stand' (Rexer 2011). Realising that taking good photographs was to either enforce a hegemony of atmosphere or to be complicit in its production led me to develop methods that might allow me to 'withdraw' from atmosphere (see Philippopoulos-Mihalopoulos 2016). Using photography as both a literal and metaphoric lens, I developed a method of approaching the OOH's built space and its situated experience through pinhole photography, which I expand upon further in Chapter 2 of this book.

I draw upon an eclectic mix of theory and literature that relates to atmospherically charged spaces (see, for example, Bille, Bjerregaard, Sørensen 2015; Ingold 2016; Degen, Melhuish, Rose, 2017; Dorrian 2014; Edensor 2015b; Gandy 2017; Philippopoulos-Mihalopoulos 2015; Thibaud 2015). In addition, I include the perspectives of artists, designers, planners, and architects who prioritise the role of the user in their work (Eliasson 2014; Hughes & Sadler 2000; Pallasmaa 2012; Sennett 2018; Sennett & Sendra 2020; Schumacher 2018; Till 2005), as well as philosophers Gernot Böhme, Peter Sloterdijk, and

Jacques Rancière who likewise are concerned with the individual's role amidst the engineering of space. Engagement with these theoretical thinkers reflects my use of atmosphere and its application as I move through the chapters. As I will unpack in more detail in Chapter 2, Böhme provides a starting point for thinking about participation through his conception of atmosphere as an 'in-between' phenomenon that includes material and immaterial elements as well as the experiencing subject. From Böhme's essentially phenomenological understanding of atmosphere, I move to engage with Sloterdijk's writings, exploring participation as occurring within homogeneous atmospheric spheres. Here, atmosphere also produces affect but is contained within a sphere of experience, suggesting a more subversive and enclosed view of participatory potential. Finally, I use Rancière's writings on the 'distribution of the sensible', and 'dissensus' (2010) to open up to an expanded notion of atmosphere[3] constituted by complex aesthetic arrangements.

In contemporary human geography and social science literature, there is a general consensus that the production of space involves 'a constellation of trajectories' and a 'throwntogetherness' (Massey 2005, 149) of entities and elements. Architecture has taken a similar 'spatial turn', incorporating the ideas of theorists such as Henri Lefebvre, to challenge the assumption that architecture is the privileged producer of space (Stanek 2012, 50). I follow this line of thought to regard architecture as an unfolding of space, 'the material engagement with person, perception and place-making' (Bille & Sørensen 2016, 237). This calls into question the notion of design as a linear process that ends with the completion of the architectural object. Asserting architecture as emergent and thinking with an atmospheric lens, I move to understand aesthetics as *aesthesis*, a term that designates the shift from seeing aesthetics as a value judgement and matter of taste to an inquiry into the conditions that stimulate relations with an object such as architecture to be 'felt and thought' (Rancière 2019b, x). As Aesthetics becomes aesthesis, I locate participation as a focal point for unfolding how meanings and perceptions of place emerge through the multiple perspectives 'between the aesthetics of reception and the aesthetics of the product or production' (Böhme 2014, 43). In this way, I follow participation as an aesthetic pursuit relating to material and subjective entwinements that are part of the ongoing process of architecture. Such 'co-fabrications' complicate the notion of user participation because they reveal how the world participates in our thinking before the decision to participate is taken (McCormack 2014, 188). Consequently, participation becomes part of a continuous and fluid process, where the invitation to take part 'colours how the engagement and interaction emerges and evolves' (McCarthy & Wright 2015, 6).

The manner in which urban development endeavours to deliver participation through immersive built environments is contrasted by the realisation that such spaces compel users to act, understand, and feel in certain ways. The staging of atmosphere becomes a central focus of this study, as it is here that a comingling of designed elements, feelings, memories, and values is played out.

Figure 1.4 Children playing in the foyer. Photograph by the author.

The Nordic invitation to participate

An invitation to participate is integral to a new Nordic approach to developing welfare-based democratic public spaces (Nielsen 2012, 178) with an 'abiding conviction [...] that the art of building is also a social art, an art fundamentally devoted to place, people, and community' (MacKeith 2008, 135). This recent move towards user participation is applied in the broader sense of bringing in new users and facilitating their active engagement (Sørensen 2016, 6), underwritten by egalitarian aims, access, and use (Valtysson 2016, 198). Consequently, some architectural voices have declared 'a new golden age' (Ibler 2014) of Nordic architecture, identifiable by its shift from designing objects to putting people first (Risom 2016, 218). The Nordic approach draws heavily upon figures such as the Danish architect Jan Gehl, who was in turn influenced by the work of activist and urbanist Jane Jacobs (see, for example, Jacobs 1961).

Gehl Architects were in fact involved in the analysis and planning of the zones around the OOH, and part of the extensive development of the greater Oslo fjord project (Gehl Architects 2007). Contracted along with Danish landscaping company SLA, they were tasked with enhancing the 'life between buildings' (Gehl 1987), designing the spaces that would conjoin the various zones of the Fjord City into a cohesive whole. Gehl's general approach offers a guiding philosophy on how humans relate to their built environments as well as providing a method for studying and implementing new spatial designs (1971, 2010, 2013). His methodology relies on counting and measuring human behaviour, laid out in a 'toolbox' of techniques that Gehl has personally developed over time (Gehl & Svarre 2013). Like Jacobs, Gehl's work is an attempt to see urban life from the perspective of the user, to install a kind of universal welfare applied to all equally, and therefore stands in opposition to the modernist view of the city as a system (Nielsen 2012, 173) where bodies are persuaded to follow a greater *plan*. However, by seeking pragmatic design solutions, Gehl attracts critique as designing for people inevitably involves hierarchies of power that often fail to 'construct meaningful consent for participants' (Pløger 2021, 2), thus resulting in the instrumental design of urban space. In other words, the call for a free and open function to dictate form can end in a fixed spatial realisation, and developments such as Oslo's Fjord City deliver a 'staged urbanism in which there is no room for irregularity and the unexpected, a neat and tedious urbanism based on a simplified understanding of the urban combined with more ideal aspirations' (Aspen 2013, 186).

In this light, the Nordic approach to development can be read as an attempt to combine progressive egalitarian intentions with a modernist planning agenda, encouraging architects to adopt increasingly, 'activist, humanitarian, and participatory practices, reinfusing the profession with a degree of social consciousness' (Weiss 2020, 235). Snøhetta tacit this approach, believing that by making a building more like a park with multiple paths and entrances, the hierarchy of architectural form can be diminished, leaving it up to people to shape their own experiences (Thorsen & Obrist 2011, 55). In addition, Snøhetta's method reveals something of 'the particularities of the Scandinavian Welfare City and its intimate relation to the practices of experience city design' (Marling, Jensen, Kiib 2007, 22). Drawing upon local and global perspectives highlights the way urban atmospheres leak, summon different scales and temporalities, and defy spatial borders.

Immaterial architecture

An obvious example of the global perspective in urban development is the recent tendency for cities to invest in 'exceptional' architecture as a means of solving cultural, economic, and social issues (Alaily-Mattar et al. 2020). It shows participation to be a contentious pursuit, often motivated by well-meaning ideals (Roberston & Simonsen 2012, 2) but applied via collaborative processes, multiple perspectives, and agendas that make it difficult to evaluate the capacity of democratic architecture to deliver democratic results (Weiss 2020, 235). That the OOH was always intended to be an iconic building is in keeping with other waterfront spaces, typified by Frank Gehry's contentious Guggenheim in Bilbao and its notorious *Bilbao effect.* Desired by city developers, often maligned by critics, the Guggenheim showcases how 'star architecture' can shape regional identity by expressing what the city is and what it wants to be (Marshall 2001, 9). On the other hand, critics decry the Guggenheim for its lack of sensitivity to the local, creating a disconnect between building and site (Foster 2003, 38). Nonetheless, inspired by its perceived economic success, cities like Oslo have attempted to emulate the Bilbao effect for area renewal, branding themselves 'while at the same time finding gainful use for a central but underutilised part of the city' (Hofseth 2008).

Combined with its original function, that of a venue for Opera and Ballet, the OOH assumes 'the role of hybrid cultural projects in the urban transformation process' (Marling, Jensen, Kiib 2009, 22). Its multifunctionality places the OOH amidst debates around the way cities can become fluid and productive by breaking down the segregation of different cultural perspectives (Kiib & Marling 2007 109), creating conditions for a more tolerant and culturally diverse 'creative class' (Florida 2002). Such arguments reveal how culture, economy, and social issues become entangled through the process of urban development which in recent times have turned to experience, particularly in Nordic lands, where sensuous interventions have become a common tactic, resulting in 'the atmospheric city' (Bille & Schwabe 2023).

The debate around how Oslo should move forward as a city was played out against these diverging perspectives of the role of architecture, art, and society. Local authorities even arranged a conference to directly discuss the impact of the Bilbao effect (Hofseth 2008, 102) showing a degree of reflection around how to best use culture in context with urban development. Such introspection was necessary for public appeasement, especially considering the elitist connotations of Opera and Ballet and other high culture projects that have given rise to a form of 'acoustic gentrification', fostering specific forms of cultural capital which alter the structure of the city (Jasper 2020, 1127). Nevertheless, city planners continue to be attracted to the idea of iconic cultural buildings as a way of resolving a number of urban issues simultaneously. Charles Jencks writes about a 'strange mood' in urban planning, 'where the architect and society both have misgivings about the iconic building but cannot help producing it, in ever greater numbers' (2006, 3). Consequently, architecture and design have seemingly internalised the critiques of instrumental planning into their thinking, creating an aesthetic not of beauty but of guilt (Koolhas quoted in Foster 2013, 122), avoiding the *mistakes* of deterministic top-down design by emphasising user experience. This is exemplified by Snøhetta's foregrounding of the individual's journey[4], wherein people explore 'self-expression, self-direction, and self-determination [...] instead of following the rules' (Boswijk et al. 2008, 45). The outcome of this is that institutions like the OOH are no longer judged in terms of their use value but by their 'ability to co-transform the sensation of the subject' (Klingmann 2007, 6). Here we can talk of architecture no longer focusing on objects but on what is between them and what surrounds them (Thibaud 2015, 41). These types of buildings blur the distinction between object and experience, using elements such as light to create sensual atmospheric encounters (Bille & Sørensen 2016). In this light, the OOH is an example of a new generation of iconic architecture, an institutional structure that espouses transparency and accountability to the public while being open to a multiplicity of users and experiences (Gandy 2019, 185).

However, the move towards an 'immaterial architecture' (Löschke 2016) brings urban development into dialogue with the 'experience economy' (Pine & Gilmore 2011) and the shift in economic output from goods and services to the commodification of experience. Companies and institutions take the role of *experience stagers* that provide intensely personal engagements (ibid., 17). According to Pine & Gilmore, it is in this way that experience, in the experience economy, becomes democratised. Individual experiences are never the same as they occur between the staged event and personal states of mind and being (ibid.), echoing philosopher Gernot Böhme's notion of atmosphere as the co-mingling of subject and object (2017, 121). Relatedly, Böhme attaches the design of atmosphere to capitalism and an 'aesthetic economy' (2003), wherein, the products of design, 'stage-set the person who buys them' (Borch 2014, 100). Unlike Pine & Gilmore, Böhme cautions against the act of designing to manipulate our unconscious (2017, 28), and despite the democratic claims for designed experience, there is concern that as cities use spectacular architecture to distinguish themselves (Ponzini 2012, 3), the new focus on experience only reintroduces the inequalities of power via building under a new flexible form of capitalism (Kaika 2011, 989). By acting as catalysts for experience, architecture can determine the way we participate in the restructuring of our urban spaces (Klingmann 2007, 2), ultimately disregarding social issues in favour of neo-liberal-driven trends (Gonzalez 2011, 1413).

A related area of contention, the focus of Chapter 7, is the use of public art in contemporary urban development projects. The issue here is how art architecture spaces make art a constituent part of the global trend of city rebranding (Miles 2015, 2). The OOH's integrated approach builds various artworks into the architecture. It speaks to a shift from individual artist-creator to 'low-key' works aimed at 'encouraging both a sense of ownership and use of public space' (Miles 1997, 122). However, using artists such as Olafur Eliasson to add immersive lighting installations puzzlingly combines a globally revered artist with the claim that art is nothing without the participation of the user (Eliasson & Kroksnes 2011, 120). The ambiguity between the status of the artist and art as a means of shaping participatory public space therefore becomes played out in the OOH's architecture and tensions abound. Hence, it is possible to simultaneously assert that the audience produces the meaning of the work through their empathy with, and recognition of, 'the overlapping connections with the world' (Rogoff 2002, 133) whilst also recognising a tendency for public art to produce mundane encounters where the staging of experience does the looking for us (Foster 2013, 209).

The structure of this book follows a composite of my many journeys through the OOH. On route, I draw upon my own experiences as points at which meanings and understandings take shape. These 'ah-ha' moments are the coming together of various other experiences: knowledge garnered from observation, explanation, and conversation. As such, my journey begins outside of the OOH before entering the building itself to explore the architecture and layers of participatory experience as they unfold. Using my own ethnographic participation as a guide I pause to address certain themes: thresholds of the OOH's space made manifest through a notion of atmosphere encountered in the field.

Chapter 3 is the first of five analytical chapters exploring different modes of participation at the OOH. Chapter 3 stands apart from the other analytical discussions, removing us from the immediacy of the building's space and tracing its historical context. It is this exposure to the OOH's development that I argue shapes people's perspectives of the OOH in situ, in what I term a form of *transformative* participation. The chapter serves to detail the case through the important elements and events that culminated in the OOH's development. These historical points reveal how a process of indirect participation unfolds, involving issues and public debates that surrounded the OOH, eventuating in specific knowledge and understandings, that, in turn, play a significant role in how local Norwegians experience the OOH. That participatory experience involves previous understandings is nuanced in Chapter 4 where I look to the OOH's *materiality*. Here I interrogate material participation by focusing on the marble that adorns the OOH's roof areas. I draw upon Gernot Böhme's atmospherically orientated reading of a material's social and synaesthetic characteristics, to unpack its ambiguity of meanings and affects. I discuss the material's qualities, such as whiteness, which serves to enhance certain participatory practices but at the expense of others.

In Chapter 5, I take the analysis of architectural experience further by concentrating on *movement*, shifting from the surfaces of things to what is played out upon and between them. I elaborate on what people do with their bodies, where and why they occupy certain areas more than others. Theoretically, I turn to Tim Ingold's ideas around *dwelling* and *inhabitation* to interpret how the OOH's architecture resonates with its users through movement. The chapter goes on to underline how a local sensibility to movement is produced via cultural understandings of Norwegian outdoor relations, played out on the OOH's roof. This calls into question the opening of experience that Ingold's writing imply, finding that a partial closure of meaning is also re-constituted through movement at the OOH. In Chapter 6, I enter the OOH's foyer to look at *light* and the immaterial aspects of the OOH already developed in the previous chapters. I describe the way natural and artificial light form ever-changing transitions, exploring how a vaguely defined understanding of Nordic light works in tension with subjective experience. Drawing upon Peter Sloterdijk's theories of *foam* and *spheres*, I argue that the implied potential for multiple atmospheres and individual agency becomes played out under an atmospherically conditioned continuum, and light is shown to be pivotal to this. In this way, the atmosphere of the OOH extends beyond its physical boundaries, constituting individual experiences that are isolated but always in homogeneous relation with one another.

The final analytical chapter looks at *Art* participation, centring on Olafur Eliasson's installation piece *The other wall* as a means of exploring the way people react and engage with public space that includes the uneasy relation between art and architecture. Analysis of *The wall* comes through Rancière's *distribution of the sensible*, and I explore how the art foregrounds specific meanings and political ideals in conjunction with the architecture, notions of process, and a tacit antagonism towards the traditional institution and its values. Moving on, I consider how the work affords individual agency through the theory of *dissensus*, as a measure of the participatory act. Underlying this chapter is the question of how social media and Instagram users participate with the art and what this means to our understanding of a notion of atmosphere. Here, I suggest that, whilst practices are varied and require nuance, the art object through digital technology asserts rigid structures that guide the user and problematise participatory experience.

In Chapter 2, a theoretical discussion of atmosphere is combined with my methodological approach and the way I have utilised an atmospheric lens in practice. I explicate how the writings of Gernot Böhme, Jaques Rancière, and Peter Sloterdijk inspired my conceptual and analytical thinking, unfolding over time and in response to my encounters with the field. It reflects upon the participation of research, its role in the construction of atmosphere, and of being *in*, thus co-constituting the object of study. I outline the interventions designed in response to the challenges of researching through an atmospheric perspective, specifically my use of pinhole photography as a means of *withdrawing* from atmosphere, as well as the notion of *dissonance* as a means of revealing our attunement. These explorative ideas are brought into alignment with other ethnographic data, highlighting the multiple perspectives and shifting positionality an atmospheric lens compels the researcher to consider and reflect upon.

Notes

1 Since 2000, Oslo has embarked on a vast redevelopment of its harbour areas, referred to as the Fjord City (Oslohavn 2021).
2 See, for example, Stenslund's study of Danish landscape-architecture practice SLA (2023).
3 Rancière does not use the term 'atmosphere', but I engage with his notion of the 'distribution of the sensible' as a means for thinking about the way meanings and practices are shaped in relation to the individual 'spectator'.
4 In conversation with Snøhetta, they referred to the journey of the user. It was said that whilst the final destination was predetermined, how the visitor arrives there was yet to be decided.

2 Adopting an atmospheric lens

This chapter discusses the challenges of utilising an atmospheric lens and the tensions that arise when conducting research whilst being immersed in atmosphere. If atmosphere is 'the medium in which we find ourselves' (Böhme 2017), then it must include the research space, compelling the researcher to reflect upon their methodological and conceptual understandings. In the field, thinking with atmosphere foregrounds their intangibility and the way they leak, making borders seem permeable. By embracing these ambiguities, we can learn to attend to their affects (see Anderson & Wylie, 2009, 78), using atmosphere to reveal how our relationship with built spaces unfolds.

Making connections between atmospheric elements bestows the researcher with an agency that should be acknowledged. The analysis that begins in the field is also born from individual sensibility and a particular ability to empathise with the expressions of those entities encountered. In a sense, this implies the participatory potential of the OOH's spatial experience, for the researcher is also a user of sorts and encounters are diverse. However, with this understanding comes the realisation that we overlook, ignore, and delete many more connections than we are able to make.

Analytical themes arrive through multiple data sets, knowledge, and experiences that extend beyond the situated object of study and into multiple times and places. Here we can think of research as a kind of atmospheric attunement, comprised of shifting perspectives and positionalities. In tracing the shifting edges and boundaries of atmosphere, I have drawn upon the writings of Gernot Böhme, Jacques Rancière, and Peter Sloterdijk, attempting to hold their ideas in dialogue with the empirical and epistemological. In the field, moving through the OOH brought about a method of visualisation, a conception of atmosphere as a *sphere* of resonance and enclosure (see Anderson & Wylie, 2009, 80), where what is inside and what is outside atmosphere requires constant navigation. The (atmos)sphere, various arrangements, and scales make the notion of a totalised system unhelpful in trying to account for the present (see Stewart 2007, 1). Moreover, thinking about the ontology of an atmosphere leads to a process of locating the hidden 'generators' (Böhme 2013, 3) where encounters in the field assemble the situated, the material and immaterial, with intentions and ideals, memories, and past experiences.

DOI: 10.4324/9781032668802-2

Figure 2.1 Thinking ontologically: A pencil sketch of co-mingling spheres. By the author.

During the summer of 2021, I made the last of my field trips to the OOH. The building looked much the same, but it *felt* somehow different. The OOH was just coming out of the pandemic lockdown and was on limited opening hours whilst no shows were being performed. There was a steady flow of people on the roof despite the cold and windy weather, yet inside the foyer the building felt subdued and under-visited. I spent several days observing and sketching, mapping the flow and behaviour of those who entered the interior space. The relative quietude gave me an opportunity to speak to members of staff who were now easily accessible and in a relaxed mode of working.

One morning, I arrived having arranged an interview with a member of staff working the front desk. When I approached her, she quietly told me that she had been told not to speak to me today. This confounded me as I had been given permission to speak openly with staff members if they were themselves available and willing. I naturally respected her wishes and moved away to sit down and assess the situation. Shortly afterwards, the security guard, who I was now familiar with, came over to me and somewhat apologetically said it had been noticed that I had been hanging around the last few days talking to people, and did I have permission to do so. I told him I did and from whom, and he said as far as he was concerned that was fine. However, when I approached other members of staff later that day, what I had previously experienced as an open and transparent atmosphere had changed and I was curtly informed they were too busy to talk. Feeling the change in atmosphere, I withdrew myself.

I relay this event to emphasise how adopting an atmospheric lens involves multiple overlapping perspectives, what Sumartojo and Pink describe as being *in*, talking *about*, and moving *through* atmospheres (2018b). The sudden rupture in atmosphere I experienced forced me to reflect upon how my role as a researcher was complicit in the very subject of my study. I had already realised how walking around the OOH, taking pictures, kneeling, reaching, and touching were the acts of yet another visitor interacting with the building in a manner that also propagates the space as a novel and unique place. But here, my own presence was directly shaping atmosphere, at least amongst some of the staff. Consequently, my object of study had become tainted with an unpleasant feeling, partly of my own making[1], and I had become a constituent of the very atmosphere I was looking to study. Unable to extract myself from atmosphere, I withdrew to the foyer's periphery and considered my position. Indeed, as this chapter will later explain, withdrawing myself from atmosphere became a tactic, a method for disrupting and thus revealing the work that atmospheres do.

The prior atmosphere of openness and transparency at the OOH had suddenly become something quite other, and the dissonance I felt actioned a withdrawal from a specific staging of atmosphere. I was pushed to its periphery, and for a time, allowed to look in. My reaction had been to move to the edges of the space with my back to the wall, a position I had often assumed when observing people's movements. However, in this new atmosphere, I was no longer *safe* at the foyer's physical edge. I had been observed observing and was now implicated in its production. Nonetheless, it disclosed that withdrawing from atmosphere stimulated an ontological perspective that supplemented the phenomenological moment, revealing the various elements involved in the staging of an atmosphere by highlighting their presence and accounting for the way they were drawn into atmospheric relations.

The atmospheric generators such as the OOH's windows, the light, and movement of people were aligned with members of staff. Had the windows narrowed slightly, and the light now dimmed. Was the space suddenly heavier, more constrained? Employees were shown to be essential to the prior atmosphere of openness, but this only became evident through rupture and the emergence of a different atmosphere (see Philippopoulos-Mihalopoulos 2016, 162).

I will pick up the themes of withdrawal and its potential to cause ruptures in atmosphere in the following analytical chapters, where I add to the discussion through the notion of *glare* and the overabundance of light, as well as *mishaps* and unforeseen bodily encounters. Relatedly, later in the chapter, I outline my use of pinhole photography as a means of disrupting the dominant visual aesthetics of the monument and its iconic architecture.

Atmospheric perspectives: Böhme, Rancière, and Sloterdijk

As I will detail in the following pages, I draw upon the work of Gernot Böhme, Jacques Rancière, and Peter Sloterdijk aligned through their contribution to the conception of atmosphere[2]. In keeping with the notion of the sphere, Böhme directs us to a notion of *in-between*, two spheres overlapping to form a third atmospheric space; Rancière, to a *distribution of the sensible*, an irregular shape that encompasses knowledge, space, time, and the senses; and Sloterdijk to *spheres* as bubbles and foam, dynamically interacting objects within a continuum.

By my reading, Böhme's writing opens the door to an expansive notion of atmosphere, a sensorial and experiential neo-phenomenological pursuit (Michels 2015, 256), involving entanglements of culture, synaesthesis, and movement (Böhme 2017, 93). Yet, Böhme also returns us to the primacy of the phenomenological moment and a notion of atmosphere that is never determinate of the experiencing subject. In fact, Böhme avoids the term *attunement* for the very reason that it implies the pre-existence of a space to be attuned to (ibid., 91). Instead, atmosphere is a vague co-presence of object and subject and the relations between an environment and human states (1993, 114). Applying such ideas in the field speaks to the experiencing individual but gives little direction when confronted with a wide spectrum of potentially determinate and indeterminate elements, arguably shrouding the power dynamics involved in large-scale development projects. However, Böhme states that the conditions we can apply to atmosphere include everything and everybody; 'there is no area of life, not a product, installation, or collection, that is not the explicit object of design' (Böhme 2014, 45). In this regard, the ontology of Böhme's atmosphere surrounds us, but can only be realised, or studied, locally through the emotional register of the subject (ibid., 46). My contention is that a designed waterfront space, despite its synaesthetic and emotive qualities, has a very different register than that of a *natural* space like a beach, and resonates with the intentions of design that should not be bracketed away from the experience of place. Here, I turn to Rancière and Sloterdijk to expand upon the primacy of the phenomenological encounter found in Böhme's conceptions of atmosphere (see also Philippopoulos-Mihalopoulos 2016), whilst maintaining the essentiality of the situated participant. Concerned with modes of participation as an *aesthetic* practice, the challenge was to retain some of the colour of individual experience, exploring the relations between things and the dispositions of the subject, whilst locating common understandings and shared encounters that might also arrive from other spaces and temporalities.

To speak of a theoretical lens suggests a fixed frame from which to interpret the world. By engaging with different perspectives, the atmospheric lens becomes unstable, a kind of theoretical kaleidoscope that holds analysis open and unfolds within the research space. And whilst Böhme and Rancière provide some instruction on how atmosphere might be studied, its elusive nature requires constant reflexivity on how it is applied to a specific space. By probing the nature of atmospheres as well as how we can come to know them, atmosphere becomes an indirect object of study that binds theory and method, which, in relation to the study of participation underscores situated experience as an entanglement spread across time and space.

There is a danger in further unsettling an already elusive concept of atmosphere, finding atmospheres everywhere and conceivably ending in redundancy. Yet, if we are always in atmosphere, then why focus solely on tangible moments of attunement instead of the mundane drift of the everyday? I take on board that, 'not all quasi-things and situations are atmospheric' (Griffero 2023, 13), but also recognise that they have the potential to be. A space that resonates with us one day, may not the next, leaving us 'alienated' (Rosa 2020, 30) but if atmospheres hold the potential for individual agency, then any assemblage of elements may constitute an attunement through its 'ordinary affects' (Stewart 2007). Hence, I take atmospheres to be ubiquitous and that we are always in a *mood* of some kind (Ahmed 2014, 14), but also that designed participation shows how attunement is never neutral, always shaped by the practices that structure our public spaces (Mouffe 2005, 806). Such a reading relaxes the grip of the individual subject, affording a more general discussion of the qualities of things and their atmospheric potential in a social context. Furthermore, it accentuates how experience is constructed as well as felt, thus contributing to the politics of perception (see Howes 2010, 335).

Consequently, I approach the space of the OOH as a lacuna between staged architecture and in-situ experience that moves across a spectrum of abstract relations. Thinking about how people participate requires being attentive to the interaction between things, humans, and nonhuman entities where, through analysis, their relations crystallise. I offer these differing perspectives not as an unnecessary apology for occupying the insecure ground of non-positivist research (Rudestam & Newton 2007, 106), but as a tentative contribution to an expanding atmospheric inquiry that recognises the importance of learning to perceive atmospheres that affect and influence us, even as they remain unnoticed (Böhme 2017, 119).

Thinking about atmospheres

'What, in the end, makes advertisements so superior to criticism? Not what the moving red neon sign says-but the fiery pool reflecting it in the asphalt' (Benjamin 1979, 86). I read Walter Benjamin's words as speaking to an atmospheric encounter with urban space, conjuring an image that moves us beyond the representational and presaging thinkers such as Philip Kotler who likened the power of atmosphere as a marketing tool to the effect of the bell on Pavlov's salivating dog (1974, 54). We apprehend the neon light but are directed to a place beyond our reason. Moreover, his quote points to our modern condition as shot through with mediated experience, leaving us unable to see with an 'innocent eye' (ibid., 85).

Theorising atmosphere in the abstract tends to become unsettled as soon as one enters the field. Considering Benjamin's motif through Böhme's conception of atmosphere, we are presented with a 'staging' of conditions: neon sign, water, asphalt, and surfaces aligned through a perceiving subject. Benjamín's atmosphere would emerge through the coming together of object and subject but result in a 'subjective fact' (Böhme 1993, 114), thereby asserting the essentiality of human experience as key to understanding atmosphere.

By referring to advertising, Benjamin suggests a covert intentionality delivered through the interaction between light and water. It shows us how a particular arrangement of the material world can set the conditions of our experience (Böhme 2013, 3), but it also highlights the excess of atmosphere and how elements like rain intermingle with staged materials in unexpected ways. That stated, and as I will discuss in the coming chapters, opening design up to the dynamic nature of the world is itself an approach, a contrivance of atmospheric design.

If research tends to commit itself to either the materiality of Benjamin's sign, or the experience of its reflection, then atmospheric theory identifies these pillars, not as separate, but relationally interconnected parts of a dynamic 'in-between' (Böhme 2017, 125). Whereas Benjamin's motif suggests a choice between pursuing the material objectivity of the neon light or the coercive *feltness* of emplaced experience, Böhme presents us with a 'staging' of conditions that seeks to communicate with the subject in specific ways. From an atmospheric perspective then, architectural staging is always in dialogue with the experiential.

It has been argued that the phenomenological method of describing the world leads research to the point where the appearance of the world is all it can know (Harman 2020, 16). By reducing atmosphere to the perception of the subject we are in danger of ending at such a point, instead of questioning how knowledge and understanding informs the felt qualities of a space, even as they remain hidden. One solution may be to take the inherent dynamism of atmospheres as 'an opportunity to look beyond the limited phenomenological opening between human consciousness and the world of things' (Philippopoulos-Mihalopoulos 2016, 155). Expanding the category of objects to include the immaterial – air, light, even histories and ideals – democratises the perceived value of things. A mountain is not merely the totality of views but the reality that makes views possible (Harman 2020, 18). Looking up at the OOH from the city, I bring its vision into the space when I arrive. But this is more than a view, it involves the knowledge that the building slopes in a certain way, designed by a certain designer, and that millions of people have made a similar journey. My understanding will differ from others, but there is always a shared part of architectural experience that involves a feeling of togetherness whilst being apart (see Thibaud 2015, 41). It allows us to transcend the notion that the mountain experienced is now different only from our perspective, pace Rosa (2020, 35), as well as making objects, both material and immaterial, a starting point of analysis. Therefore by, 'taking the autonomy of objects seriously, [we] can begin to investigate relations between non-human objects without reducing these relations to how they appear to human beings' (Ash & Simpson 2016, 60).

An observance to Böhme's ideas leads us to prioritise the phenomenological, focusing on the perception of the 'fiery pool' and neglecting the outside world. What is lacking is attention to the positioning of the lights, the choice of lettering, the decision to use a certain hue of red, and the intentions of designers and planners who have arranged space to communicate with a certain type of *user*. In other words, within the context of designed spaces, an atmospheric perspective should not overlook the invitation to participate as part of the feltness of space. Consequently, following the ontologically indistinct can be the foundation for understanding the connections between tangible and intangible aspects of human life (see Bille & Sørensen 2016, 3). This is not to disconnect participatory experience from the phenomenological, but to plumb its limitations and potentials, exploring how atmospheres become 'composed and re-composed' (Michels 2015, 258), designed and experienced from an expanded view of the agency of things in the world.

Interactions between entities become meaningful occurrences central to human experience, thus broadening the ontological perspective from which analysis can be undertaken. Enforcing a self-reflexivity to one's own interpretations as part of the research process (Morton 2018, 27) reveals the intersubjectivity of subjectivity and the way we are bound to all manner of elements, disrupting the assumption that we can only approach the world through the subject (Anderson & Wylie, 2009, 323).

Taking photographs was an important part of my own ethnographic practice, but during fieldwork it became clear that another form of image-taking was an important part of people's participation at the OOH. People were taking their own images of place and self, primarily on mobile devices, the results of which could be displayed and viewed on social media platforms. This leads us to the multimodality of contemporary designed spaces, particularly in the realisation that digital media produces and distributes images that alter the way a space feels. In this sense, digital media becomes a constituent of atmosphere, part of a 'new urban aesthetic' that raises questions about how urban experience is being mediated (Degen & Rose, 2022, 5).

The act of imaging and editing software is perhaps beyond what Benjamin could have envisaged, which makes us ask whether digital media presents an entirely new aspect of atmospheric staging, or just another incarnation of the tensions between desire and rational thought. Regardless, by attending to the multimodal nature of the OOH, technology such as mobile devices are shown to have profound implications for the phenomenological experience of urban space and architecture (see Degen et al. 2017; Degen & Rose 2022). In the context of this book, it forms a layer of participation inseparable from the analytical themes discussed. From an atmospheric perspective, how social media allows people to be simultaneously situated and potentially absent, further challenges the primacy of the phenomenological moment. It suggests that atmospheres be experienced from afar, from outside of the immediate space.

Rather than holding on to atmosphere as a situated quasi-object phenomenon, the multimodality that comes with social media demonstrates how atmospheres creep beyond the borders of staged space. Or rather, that a notion of staging must also include technologies and perspectives *generated* elsewhere, that are different but inseparable from the experience of situated space. Furthermore, through digital media the expression of the built can undergo significant modifications, enforcing certain perspectives whilst backgrounding others. In turn, this leads designers to seek to control the mediation of their designs, pre-empting what people will do, as is the case with the *selfie* wall in architecture (see MacDowall & Budge 2021).

Ultimately, what multimodality highlights is how we should be observant of the way atmospheres are subjective yet interpersonal and shared. Therefore, the challenge is not to divide Benjamin's motif into parts but to tell the story of an expanded range of elements and how they converge to shape particular atmospheres.

Building experience: architecture and atmosphere

Keeping Benjamin's motif in mind, here I wish to outline some of the theoretical ideas of Böhme, Rancière, and Sloterdijk in relation to atmosphere and the design of spatial experience. Leading me to these specific thinkers was the question of atmospheric design in relation to its participatory potential. It involves a design perspective which implies atmosphere to be a tool for enhancing individual experience and its counter position that atmosphere is a coercive method that enforces particular perspectives and ideals through the shaping of experience.

Within architectural practice, there is an increasing recognition that the resonance of a space is fundamental to a building's success and that built environments are able to channel immediate affective responses (Adey 2008, 441). Moreover, this design perspective infers that enhancing individual experience is to democratise a space, endowing it with participatory potential. In so doing, atmosphere carries emancipatory ideals but also the logic of capital and the experience economy that seeks to brand locations in specific ways.

Böhme uses the term aesthetic labour to define the activity, 'which aims to give appearance to things and people, cities and landscapes, to endow them with an aura' (2003, 72). This Böhme associates with the 'aesthetic economy', a new phase of capitalism which has moved from creating objects of use value, to the endowment of things with an appearance that stages and intensifies life (ibid.). Here atmosphere speaks to the way users participate with urban spaces in sensuous and bodily ways, through the immaterial characteristics of the built.

By turning to situated experience and the immaterial, architecture tacitly addresses the perceived failings of modernism. Frank Lyons links architectural form to spiritual and ineffable 'nothingness', noting that immaterial architecture opens buildings up to interpretation, inserting an incompleteness that resists the precision of Euclidian geometry (2019, 95). From this perspective, architectural atmospheres are emancipatory, bestowing the visitor with agency as it, 'stimulates activities and guides the imagination' (Pallasmaa 2014, 19). It follows that designers are increasingly pursuing the immaterial and the way elements like light and sound shape space experientially (Böhme 2016). For example, architect Peter Zumthor defines his work as informed by the felt and emotionally sensed (2006). Echoing Böhme, Zumthor writes about atmospheric experience as drawing materiality and subjectivity together. Similarly, Malcolm McCullough conceives of atmospheres as a 'fixity', and architecture as the spatialising entity that holds together eclectic meanings by way of its mise-en-scéne (2013, 173).

Unsurprisingly, architects tend to emphasise the importance of design and atmosphere as tools for affording user experiences that, 'unavoidable in terms of perception, should therefore always be part of the design process' (Weidinger 2018, 13). However, making the agency of the designer explicit reveals their power, even as they might state their intent to hand it over to the user. As curators their task becomes to afford us new and diverse encounters, turning our cities into places where people can experience synesthetically by prioritising multisensory consciousness (Pérez-Gómez 2017, 15).

This focus on the user and their perception of the environment generally overlooks the relationship between designer, user, and design. Juhani Pallasmaa, an architect and writer who places great emphasis on the haptic potential of architecture writes, 'we enter a space, the space enters us, and the experience is essentially an exchange and fusion of the object and the subject' (2014, 20), obscuring the practice of design in favour of the situated encounter. Relatedly, urbanist Richard Sennett has written extensively on the qualitative nature of experience through his use of the terms *ville* and *cíte*, framing the difference between building and dwelling, design, and experience, arguing the need for urban development to focus on the user by affording a greater diversity of experience (2018, 5). Sennett's work relates to the tension in 'designing disorder' (2020), taking its point of departure from a critique of modernist determinism, arguing that development approaches have the potential to shape *open* experience, whilst acknowledging that architecture unchecked often results in bounded enclosures.

This tension between designed emancipation and hegemonic control is often discussed against the contentious philosophical thinking of Martin Heidegger. In 'Building Dwelling Thinking' (1971), Heidegger argues that building is fundamental to our meaningful relationship with the world, warning that aspects of modernity undermine the essence of our attunement to it, what he terms 'dwelling'. Heidegger's writes,

> when we speak of man and space, it sounds as though man stood on one side, space on the other. Yet space is not something that faces man. It is neither an external object nor an inner experience.
>
> (6)

The notion of dwelling speaks to atmosphere and its capacity to connect us to the world, and in this sense, it advocates forms of architecture that attune an audience to meaningful experience. Importantly, dwelling also involves the designer, as Ingold states, 'only because they [builders] already dwell therein can they think the thoughts they do' (2000, 186). Because the designer's dwelling shapes the form of the built, the danger is that architectural vision may fail to cater for difference, 'mixed forms and uses invite mixed uses [..] the more form becomes simple, clear, and distinct, the more it defines who belongs there and who doesn't' (Sennett 2018, 129). This is a legacy of Heidegger's thinking: a resonance with place that leads us to being concerned with what can be deemed 'authentic' (see Malpas 2021, 52) and what, or who, should be placed outside of the interior of dwelling.

Despite the agency bequeathed by the designer, Böhme's atmosphere 'remain[s] reliant on the experiencing subject, that is, human beings in their bodily presence. The space of bodily sensing – a sensing that reaches out into the indeterminate expanse' (Böhme 2017, 75). Only through the sensing human subject can the potential for diverse experience be realised, and designed space would lie dormant, unknowable to us. This is somewhat puzzling in light of Böhme's parallel emphasis on the 'ecstasy of things', 'and the way things go forth from themselves' (1993, 121). Böhme attempts to bring together the poles of subject and object, presenting us with atmosphere as an *in-between* phenomenon which is paradoxically both designable and yet undetermined by design. Atmospheres emanate from gatherings of, 'things and persons' experienced by us as something. 'quasi-objective' (Böhme & Borch 2014, 43).

This emphasis on the potential for atmospheres to always be *more* than their design or their experience (Anderson & Wylie, 2009) is grounded in the idea of the world as always emerging, ripe with potential. Here we come to a notion of atmosphere as more than felt, as a phenomenon of the natural world. For example, Tim Ingold retains the meteorological dimension of atmosphere, foregrounding weather as an essential constituent of our experience with the world (2007). Using 'respiration', Ingold exemplifies our environmental immersion as we breathe in to receive an 'invasion of consciousness', breathing out, 'in a moment of perception' (2016, 12). It is in this 'fission/fusion' action that we perceive the world, in a rhythmic cycle of 'absorption', followed by a 'propulsion' into the world (ibid.).

In addressing our entangled relations with the world, Ingold summons Heidegger's concept of dwelling, carefully avoiding its association with exclusionary building practices. The difficulty for research becomes how phenomenologically grounded thinking can avoid falling into a universalist perspective that overlooks the 'aesthetics of power' (Degen, Melhuish, Rose 2017, 7). As David Howes writes, 'sensing is public', and as researchers we should attend to 'the sociality of sensation and the cultural mediation of perception' (2022, 3). It is towards the theme of power that we now turn, as I look at atmosphere as a coercively staged experience.

Coercive atmosphere

Walter Benjamin famously conducted a study of the Paris arcades, interrogating the effects of capitalism on the intimate areas of everyday life. He saw how the outside was being brought into an interior space, how the arcades were 'covered but not enclosed', its walls serving only to partition the commercial space within (Benjamin 1999, 158). Benjamin's observation alludes to how architecture and its arrangement of materials has its own force, and how art and politics share a capacity to 'organize a common world of self-evident facts of sensory perception' (Rockhill 2009, 196).

Peter Sloterdijk highlights Benjamin's project before turning to another example of partitioned space, London's 18th-century exhibition hall, the Crystal Palace. He relays how, after visiting the palace, author Fyodor Dostoyevsky was affected to write that western society was enclosing itself in a post-historical space where a state of inauthenticity, combined with the goal of happiness, was being performed by individuals. Dostoyevsky concluded that such architecture facilitates the inevitable collapse of society, wherein 'everyone made themselves their own specialty' (Sloterdijk 2014, 154). Sloterdijk's conception of atmosphere as *sphere* comes in the form of *globes, bubbles*, and *foam*, making a distinction between inside and outside at various measures. The Earth itself is conceived as an 'absolute island' because there is literally nowhere else to inhabit (Gandy 2017, 361), whilst modern architecture forms 'atmospheric islands' (Sloterdijk 2016, 315) that involve a subtle conditioning of the air inside. Reminiscent of Böhme's *staging* of atmosphere, the air inside a Sloterdijk sphere is 'engineered' in ways that mean it is never neutral (Philippopoulos-Mihalopoulos 2016). Bubbles are spheres that contain dominant ideologies such as religion or politics. Their function is to offer a common understanding of the world by offering its inhabitants 'immunity' (Borch 2014, 65) and a feeling of protection from the outside. Here Sloterdijk asserts the importance of architecture regarding contemporary buildings as 'new event environments' (2016, 624), not mere curiosities but forms of integrating urban space.

If the general belief of designers and architects is that atmosphere has the potential to produce democratic and emancipatory spaces, then there is also a growing awareness that staging atmospheres is to exercise a form of power. In the social sciences, recognising atmospheres as a partially engineered phenomenon has unlocked interest in an array of immaterial elements such as: light (Bille 2015; Edensor 2015a), sound (Jasper 2020; Riedel 2020), smell (Stenslund 2015), air (Adey 2015, McCormack 2018), or sensorial combinations (Ebbensgaard 2017; Sumartojo & Pink 2018b; Schwabe 2021). These studies tend to analyse spatial designs through people's embodied and sensorial engagement, zooming in on everyday experiences whilst acknowledging the design process and its atmospheric staging. Often, the mode or act of participation tends to fluctuate between the individual's part in an 'experience economy', and their liberation from the tenets of modernist building. Moreover, as atmosphere enables a rehabilitation of the senses typically dominated by sight

(Albertsen 2019, 4), the notion of how spectatorship becomes enabled and performed becomes fundamental to navigating this tension.

The philosophy of Jacques Rancière does not speak specifically to atmosphere, however, his ideas provide a conduit for thinking about participation atmospherically. In his writings on the 'distribution of the sensible' (2013), we are referred to a sensory apprehension of the world that guides our interactions through 'what is visible and invisible, sayable and unsayable, audible and inaudible, thinkable and unthinkable' (McCarthy & Wright 2015, 41). The distribution of the sensible indicates a notion of atmosphere in that it highlights the fabric of experience as spread over space and time, forging the activities that 'determine the very manner in which something in common lends itself to participation and in what way various individuals have a part in this distribution' (Rancière 2013, 7). In this way, Rancière's ideas are a stimulating cohort to the more situated and phenomenological ideas put forward by Böhme. His fabric of the sensible conjures the idea of an encompassing aesthetic that powerfully shapes our perception, but this opens a door to possible ruptures and social changes that can be made within hegemonic relations.

Through Benjamin's motif, atmosphere would have the potential to shape cultural values, desires, and perceptions. In his discussion of the emotional power of advertising, most famously his analysis of Nazi propaganda, he points out the way politics has become aestheticised. Building on this, Böhme argues that aesthetics has now penetrated 'previously non-politicised fields and objects' (Borch 2014, 72). There is common ground between Böhme's 'staged' space and Rancière's fabric of the sensible as both infer potentially coercive arrangements of space, vulnerable to rupture and new realisations through the perceiving subject. Relatedly, Sloterdijk emphasises the material structures within which we are conditioned and the importance of architecture as the physical nature of the spheres that constitute the felt immunity of a space (Borch 2014, 66). It is architecture that always shapes the material membranes that give us immunity, not simply to keep danger away by reinforcing the walls, but by shaping feelings within. And, whilst Rancière writes primarily about art, he acknowledges that his ideas relate to all forms of design that divide communal space by drawing lines, arranging words, or distributing surfaces (2019b, 92).

As a result, the designers of atmosphere wield significant power as they can intentionally shape moods and emotions (Böhme, 2013, 39). The activity of staging, therefore, carries a 'double character', an empowering and life-enhancing creation of experience, as well as a potentially manipulative, even exploitative method (Albertsen 2016, 576). This double character speaks to Böhme's conception of atmospheric emergence that holds atmosphere not to be a thing, 'but rather a hovering in-between' (Böhme 2017, 161). In this sense, architecture becomes an artistic practice with the potential to, 'intervene in the general distribution of ways of doing and making as in the relationships they maintain to modes of being and forms of visibility' (Rancière 2013, 8).

Rancière also draws attention to how the narratives and form of traditional art mediums such as painting have turned to focus on 'movement, light and gaze, building its own domain by blurring the specificities that define the arts' (2019a, X). The relationship between today's art and architecture exemplifies this, underlining the immaterial and atmospheric sensibility of contemporary spaces. Furthermore, like Böhme, this leads Rancière to discuss aesthetics as *aisthesis*, the mode of experience that includes perceptions and emotions (ibid). From a participation perspective, Rancière is empowering because the distribution of the sensible points to a meeting point between the multitude of *surfaces* comprising a sensible, which, when received by our sensory apparatus, results in a form of individual agency set in motion. As Citton puts it, 'a window opens during which "the real action" can take place' (2009, 123). Importantly, agency is not bestowed upon or given to the individual but is, in fact, the innate state of affairs whenever we are confronted with a work. We always have the capacity to re-work the sensible. This recalls Böhme's assurance that individual experience can never be fully determined by design. However, whereas Böhme adopts a broadly phenomenological stance wherein agency arises through situated atmospheric encounter, an equality of relations is Rancière's radical starting point (Rockhill 2009, 108). When we participate with architecture, we partake in a reconfiguration of the sensible which is always political in the sense that it is here we come to understand ourselves as subjects within a particular context (Citton 2009, 123). Therefore, it is the 'entirely material conditions' that make it possible for an artwork or design to be felt and thought (Rancière 2019b, x). In turn, this shows urban development as the enforcement of a particular sensible, which for Rancière is the work of the establishment against the citizens' position within a society and their ability to remain unsubordinated by it (Žižek 2013, 66).

Whilst Rancière adopts a horizontal perspective, locating combinations between systems of possibilities (Rancière 2013, 46), Sloterdijk identifies the spatial configurations that form atmosphere attunement across various different scales. This scaling up from the situated to the global and beyond, from the mono-sphere to the dynamic gathering of isolated bubbles speaks well to participation in the context of urban development. As with Rancière's notion of the sensible, Sloterdijk's bubble sphere offers a means of understanding the enclosure of meaning, but Sloterdijk does so by defining edges and boundaries, showing us how they are engineered to become more or less permeable.

The most recent development of Sloterdijk's sphere paradigm is *foam* which consists of the chaotic assemblage of bubbles that 'however much they might purport to be connected with other and outside things…are drawn into an interactive network based on the principle of co-isolation' (Sloterdijk's 2016, 56). Sloterdijk gives the example of drivers on a motorway, moving with a reactive resonance, but being relationally isolated (ibid.). As with visitors to the OOH's roof, there is an attunement that comes 'not through direct exchange' but through the 'mimetic infiltration' of each isolated sphere (ibid.).

The concept of foam is particularly stimulating in light of the vagueness of Böhme's notion of inbetweenness, allowing us to conceive of bodies interacting within a space that is open, yet bounded.

On this theme, Andreas Philippopoulos-Mihalopoulos draws attention to the work of artist Tomás Saraceno and his use of spheres, 'resembling miniature glasshouses' that allow us to observe the two sides of the air. We onlookers are also a part of the atmosphere; 'we are not in it, yet we are part of the installation' (2016, 158). With Foam, Sloterdijk posits a fragmented social space, giving us an image of independent spheres in indirect relation with one another. It presents us with the idea that being in the world 'is simply being in spheres' (Schinkel & Noordegraaf-Eelens 2011, 29), but that the membranes that divide us with other spheres are made to be more or less porous. By my reading, Sloterdijk challenges our assumptions of what it means to be in a space *participating*, by providing an ontological dimension to the theory of atmosphere. Crucially, he reveals the power structures involved in filling spheres with atmosphere, conditioning the very air, which is an act of design intent on manipulating those on the inside (ibid., 30).

Research atmospheres: methods and tools

Before the Corona pandemic infringed itself upon us, I had visited Oslo on three separate occasions, each time for around a week. A fourth field trip was eventually possible in the window of summer 2020, before things again took a turn for the worse. Finally, I was able to visit in the winter of 2021 when the building was open in a limited capacity. The pandemic forced me to adapt, along with the rest of the world, in ways unforeseen. As well as talking to experts and visitors, drawing, photographing, and mapping the OOH's public spaces, it also led me to focus more on the multimodal practices I had witnessed unfolding at the OOH as essential to accounting for urban space, heavily mediated by digital media (see Degen & Rose 2022, 3).

As researchers participating with atmosphere in the field, we seek common realisation, moments where things align and make sense, and we are able to understand a subject's world view. This process of realisation is itself an attunement to atmosphere (Engels-Schwarzpaul 2017, 1), the point at which object meets subject and atmosphere emerges. When Böhme highlights how spaces can be 'tuned' (2013, 5) by designers that set the conditions, he underlines architecture as a spatial practice best experienced through the sensuous body (Engels-Schwarzpaul 2017, 3). Thus, we come to know something about the space we are in when we sense where we are, which also means being able to sense the space through our attunement to it (Böhme 2017, 74). For Böhme, this is not about determining the things and objects that exist in a space, but instead being concerned with how 'they radiate outwards into space', the way things step outside of themselves (2013, 5). It is this focus on the tone of things and not their material properties that leads Böhme to speak of the ecstasy of form (ibid.). In terms of the OOH, it is the immateriality of materials that relates to my interest in the way contemporary architecture presents evocative surfaces in order to shape sensory experiences for the user. According to Böhme, the experience of a thing that goes forth from itself holds characteristics that are socially and synesthetically defined (2017, 60), telling us that part of our reception of the immaterial is predefined through cultural understandings, such as the interaction of warm light and wood perceived as Nordic. What remains of the reception of the immaterial must be left to the individual subject and their own sensibility to create novel connections. It is in this excess that subjectivity is freed from design, and the potential for atmosphere to liberate the individual is underlined.

Concerning methodology, it is only Rancière's later writing that gives instruction on how we may begin to unpack the workings of the sensible. For him, surfaces are distributed designs that apply to the material and conceptual. Meanings arrive through mixed arrangements of materials and technologies that deny their 'autonomous purity' as they are laid out on a shared surface 'common to art and non-art' (2019b, 105). To analyse these 'melanges', Rancière suggests we follow the links between things, not as finite meanings, but as ways of revealing how surfaces are made as singular events whilst exploring the 'interpretive network that gives it meaning' (2019, xi). Designed works, therefore, carry the presupposition of cause and effect, grounded in an inegalitarian principle (Rancière 2011, 14). Yet, whereas Böhme posits atmosphere as the comingling of subject and place, Rancière argues the sensible arrives not from the reception of a work but from 'an ordered set of relations between what can be seen and what can be said, knowledge, action, activity, and passivity' (2009, 21), developing an expansive framing of the conditions that *allow* us to perceive (2019, x). With these differing conceptions in mind, participatory spaces can be approached from a situated perspective that includes *surfaces* beyond the architecture's physical space.

As a research practice, spending time by walking through the OOH's architecture was something I did often and on every occasion. Inspired by Sloterdijk's ontological approach, the boundaries and soft transitions between atmospheres were sought. Fieldwork was spread over different seasons, enabling me to observe and experience how people's activities would change in various climactic conditions. Moving around the OOH's space was also an essential means of generating knowledge concerned with participation, where walking itself becomes a form of ethnographic practice (Pink 2008). Walking the roof involved noting the coming together of imaginary landscapes and people's bodies, realising their 'intersecting emotional, mental and geographical topographies, form [as] a series of drifting chords' (Brünner 2015, 178). This inevitably involves the sensations of the researcher, aligned to others through observing and participating with others as they move around the space. In addition to bodily sensations, paying attention to the aesthetics of our own contemporary condition (Rendell 2002, 26) is important to understanding how precisely atmosphere is being staged. Movement as a 'generator' of atmosphere (see Böhme 2017, 23) involves bodies in particular spatial contexts, but it also relates to the designed intentions of the built, which, as we will discuss later in this book, produce dynamic variations through phenomena such as light and weather.

Spending time at the OOH also involved methods such as sketching, tracing, and counting. The latter can create understandings of how a space is used in relation to specific times and conditions, the former a means of absorbing the complex forms and textures of architecture. As with photography, sketching can create representations, but the act of drawing can also develop, 'becoming rather than being' (Berger quoted in Brünner 2015, 184), mimetic renderings that have unfolded over a duration of time. Really spending time to look and translate through the process of sketching is a form of embodiment that can forge imaginative connections between individual and architecture. Final renderings hold 'echoes of the reflective impulses within my own body, as much as they were records of what was observed' (Turner 2018, 38). These various approaches were combined in pursuit of an atmospheric perspective that would draw together producer, perceiver, and researcher in 'imaginative moments' (Rauh 2018, 56).

Figure 2.2 Learning to look for the elements of atmospheric. A fieldwork sketch of the OOH's roof. By the author.

In considering the aspects of design, I visited Snøhetta's studio in Oslo on two occasions and conducted extensive interviews with an architect directly involved with the OOH project. Furthermore, I interviewed several artists who had contributed to the design of the OOH's *artwork*, namely the marble roof and the aluminium cladding. Additionally, I spoke to various members of staff and key players involved in the original development of the Bjørvika area. Tour guides were a particularly interesting source of knowledge in that they combined technical know-how with personal reflections grounded in their ongoing relations with a visiting public. Moreover, as representatives of the house, guides aid the circulation of particular narratives and themes that span design ideas, place marketing, and user experience. It should be added that, in this respect, I found the OOH to be shot through with theoretical ideas. Architects and artists are often well-versed conceptually and the presence of conceptual ideas was often embedded into the explanations of the building's material form. Guides were found to function as a mouthpiece for architectural thinking, relaying its ideas to visitors that, in turn, co-constitute people's perceptions.

To explore the different registers of the user perspective, I approached visitors in situ, inquiring into their individual observations. These on-site interviews were aimed at shedding light on what users do, how they feel, and what they understand. This particular practice used a bipolar emotional response test (BERT)[3] in combination with the semi-structured interview, allowing an intuitive response to each particular participant[4]. Finally, these methods were inevitably brought into dialogue with my own aesthetic, embodied, and sensual experiencing of the site (Warren 2012, 114) allowing imaginative interpretations to be formed through 'subjective facts'. Being in and thinking about atmosphere lays some claim to being able to say something about it, whilst being attentive to my own haptic experiences aided the analytical process already begun in the field. In this way, it was possible to make *a* sense of the OOH's modes of participation as they began to surface in alignment with data from other subjects; in fact, it was crucial to the identification of the modes that now comprise the analytical chapters of this book.

Withdrawing from atmosphere: taking photography seriously

Finally, I would like to discuss atmospheric withdrawal in relation to my use of photography. The method unfolded during fieldwork in response to ongoing analysis and an ontological atmospheric approach. As a method, withdrawal can be seen in the context of atmospheric dissonance, a means of rupturing and revealing the workings of staged attunement. In the following chapters, I utilise other forms of dissonance in addition to photography where participatory modes such as movement are explored through 'mishaps' such as tripping and stumbling (see Vergunst 2016) on the roof of the OOH.

I began using photography for documentation and began the simple act of taking pictures as I walked around familiarising myself with the site. Already acquainted with the techniques of picture-taking and armed with a steady tripod, I attempted to capture *good* images for analysis and dissemination. The camera is a standard tool in ethnographic fieldwork, used to either gather or elicit data from research participants (Given 2012, 621). Despite this, photography carries the suspicion of 'lazy ethnographic' practice (Warren 2012, 3). Perhaps because the camera is a medium 'wedded to the world through the mechanism of recorded light' (Rexer 2009, 15), it has come to be regarded as little more than a passive eye. However, if we consider photography not as a window to the world but as a 'created perspective' (Leavy 2015, 224), then the type of images we take as researchers become important objects for reflection; ethnographic practices such as taking photographs can play into the dominant aesthetics of development, making research complicit in the work of the 'brandscape' (see Klingmann 2007), and the staging of atmosphere.

Exploring movement and the ephemeral qualities of the building, I made time-lapse and hyper-lapse[5] studies from various points around the OOH's architecture. Time-lapse is particularly effective when zoomed out to capture vistas against the changing background of light and weather, whilst hyper-lapse involves a situated practice that follows the movements of the image maker. Having built an archive of images that could be used for analysis, my engagement with the OOH through photography took a more experimental turn. Guided by the notion of withdrawing from atmosphere, and in response to the challenges of researching an iconic building awash with visual images and mediations, photography became a means of further developing the exploration of atmosphere's permeability.

The ubiquity of images circulated through tourism, marketing, and branding carries its own force, particularly in relation to people's participation with architecture, such as the OOH, where the 'tourist gaze' becomes organised via dominant visual constructions (Urry & Larsen 2011, 18). And, whilst photography is assumed to accurately document its objects of interest (Rose 2016, 310), photography's unique relationship with architecture complicates its use as a methodological tool. Here I am referring to the eagerness with which architects have associated themselves with iconic photographers throughout the 20th century. It raises questions about the role of photography as part of a 'systems of communication that are the true site within which modern architecture is produced' (Colomina 1996, 14), and furthermore, how photography is complicit in constituting participatory modes. From this perspective, photography becomes an intrinsic part of the staging of atmosphere. For example, a study by Marc Goodwin (2016, xi) shows how architectural photography reduces the world of atmospheric variety to 'the blue and white of eternal summer'. Böhme makes a similar observation noting how architectural photography presents 'eternally favourable weather offering ideal lighting, nothing indicative of season is visible, and there are no chance scenes' (2016, 155). Böhme goes on to point out that the modern architect is an entrepreneur, identified through his work, making its dissemination essential to their practice (ibid., 149). Therefore, by taking good photographs, I was perpetuating a distribution of the sensible and closing-off of the potential for different interpretations. The white roof, surrounded by blue skies and dotted with a multitude of diverse visitors, perpetuated the dominance of a certain architectural aesthetic whilst deleting others.

My response was to employ pinhole photography and embrace the 'slippages' (Bell 2009, 162), the unforeseen events and technical errors that can produce 'decisive moments', Henri Cartier-Bresson's term for the 'point at which action and situation coalesce' (Campbell 2021, 130). Pinhole photography compels a different type of photographic practice, emphasising the act of making images as much as the images themselves. Consequently, pinholes enabled me to turn a visually dominant medium into a more sensorial and situated ethnographic practice, that also develops the presence of the researcher in the image as it draws the viewer into imagining themselves being there (Warren 2012, 5). Unlike the distance produced in a good photo, pinholes carry a sense of proximity, of the ethnographer's experiences informed by temporalities, rhythmical recurrences, and the duration of events (see Vannini 2019, 78). In this sense, photography becomes a co-creative practice which also reveals the potential for intervention and participation in its ongoing production.

Pinholes and fuzzy

Good photography enforces certain views of the world and is complicit in the staging of atmosphere. In contrast, *fuzziness* is a term that suggests a lack of visual clarity and a general disruption of photographic conventions such as composition, light, and colour balance that liberates 'the gaze from the tyranny of the perfect image' (Hüppauf & Wulf 2009, 251). Here we might return to Rancière's ideas and note how the practice of image-making through pinhole photography can constitute an intervention in the sensible, underlining the potential for new configurations to be made, which is in contrast to the act of research pursuing the good image.

Pinhole cameras resemble the earliest forms of photography that have developed from blurred, to clarity, and then to digital (Rexer 2009, 9). Taking the extremes of photographic technology, I discarded the modern lens for a homemade aperture that throws light directly onto the digital sensor. A tiny aperture requires long exposures resulting in movement, either from camera or subject, that renders streaks and blurs onto the image. Even when using a tripod, motion blur is inevitable as the world rotates, sways, and vibrates around us. Moreover, the physics of photography dictate that small apertures increase the depth of field, which is to say, objects in a scene will appear in focus regardless of their distance from the camera. Hence, everything will appear equally, but not exactly, in focus. Furthermore, a pinhole image cannot be accurately framed, leaving the operator to approximate the scene before them. By playing with the conventions of photography, a set of dogma was enforced that instigated a 'flat way' of seeing the world (Latham & McCormack 2009, 255), pushing against dominant aesthetics.

Figure 2.3 A pinhole image showing people moving around the OOH. Photograph by the author.

Abandoning my tripod, I was forced to crouch, kneel, and lean in order to capture an image of sufficient quality. This resulted in a bodily practice in direct contact with the materials of the building as I navigated its variable planes and features. In a sense, the material structure of the OOH became a foundation, a tripod that transcended the distancing effect of the conventional image, bringing my body into relation with architectural surfaces via the camera, and produced *atmospheric objects*, images that showed the dynamic interaction of research, building, and environment.

I found further inspiration in the work of artists like Hiroshi Sugimoto whose architectural images blur out detail leaving only light and shadow. Likewise, Blees Luxemburg's work was instructive as it revealed, through dramatic exaggeration, light's role in generating atmospheres, showing architecture to be the scaffold from which atmosphere emerges (Ingold 2016, 163). Pinholes make clear where the dominant sources of light are, underlining their luminosity and relation to materials in ways that good photos tend to smooth away. In this manner, pinhole images can reveal the way atmospheres condition the interior of a space through interactions between immaterial elements and space, its physical objects, and tectonics. By stimulating an ontological perspective, the pinhole image can highlight the permeable boundaries between an atmosphere's inside and outside.

Figure 2.4 A pinhole image showing natural light interaction with the foyer space. Photograph by the author.

Böhme tells us that good photographs 'should convey experience using the elements responsible for creating a buildings atmosphere' (2016, 153), implying photography's role in locating atmospheric *generators*. Considering how elements like light are organised within architecture helps us understand the way a space is being staged. In turn, fuzziness generates its own sense of atmosphere through the space's emerging aura that can be passed on to the viewer (Baltzer 2004, 99). Pinholes present architectural forms devoid of detail, meaning they can be used to show that, in atmospheric terms, what 'matters are their radiance, their impressions, the suggestions of motion' (Böhme & Borch 2014, 45). If good photography directs the viewer by controlling light, composition, focus, and so on, pinhole photography avoids such embellishment, enabling the uncertain to be fleetingly held in place (see Rendell 2002, 26). There is a feeling of being present in the pictorial space where *fuzzy* introduces an intensity to the act of looking (ibid, 18), and an imaginative dimension that engages the viewer. As with painting, pinhole images refer to something beneath the surface of the image, 'recognized only in outline and in shadowy hints' (Baltzer 2004, 103). From this perspective, we might talk of the pinhole image as a rupture in the atmospheric design of a space, reminding us of the agency which Böhme bestows upon the experiencing subject. The dominant aesthetic of architectural photography becomes suspended through withdrawal, showing it to be just another instance of our experience and a particular distribution of the sensible.

Roland Barthes has written the photographic image is 'pure contingency' (1981, 28); it always points to the conditions of its production. A photograph always carries intention, presenting a specific image signals a desire to communicate something in particular. The tension between sharp and fuzzy photos introduces a comprehension, 'through a logic of difference' (Hüppauf & Wulf 2009, 233), where a fuzzy image is perceived as a failed attempt at creating a good image. However, when fuzziness is deemed as deliberate, it breaks with 'the logic of mimetic representation' (ibid., 245). We should therefore make a distinction between fuzzy produced through mechanical glitches or technical incompetence, and fuzzy as a deliberately produced image (Hüppauf & Wulf 2009, 234). Fuzzy, understood to be intentional, deflects the suspicion of incompetence and foregrounds the researcher/photographer, just as a series of abstract paintings hanging in a gallery '[..] so closely links the flow of paint to an artist's behaviour, that the latter seems to be the image of the former' (Bourriaud 2002, 41).

Likewise, the pinhole signals the actions of the researcher, telling the viewer something about the ethnographic practice and experience in the field as an event, what Sarah Pink has called the 'ethnographic place' (2015, 48), foregrounding how the researcher constitutes meaning, 'through their participation' wherein the reader is given the invitation to imagine themselves in the place of the researcher (ibid., 49). Following this line of thought, pinholes can prise open the pictorial space of the photograph, allowing the viewer to participate imaginatively, even experiencing atmosphere indirectly, and when not spatially present (Böhme 2016, 154), forging an opening for the viewer to imaginatively participate with the image. In this way, fuzziness can forge openings that assert the potential of situated experience, allowing a withdrawal from photography's dominant aesthetics that both reveals and shapes the atmospheric qualities of a space.

Having outlined some of the theoretical and methodological approaches adopted in the use of an atmospheric lens, this book will move to contextualise the Oslo Opera House and its role within the development of Bjørvika and Oslo's Fjord City project, against which, it is argued, a type of transformative participation unfolds.

Notes

1 It should be stressed that this was a one-off experience, occurring at a strange coming-out-of-lockdown period. The institution and its employees were otherwise extremely welcoming and helpful in their dealings with me, contributing to an atmosphere of openness and transparency.

2 Unlike Böhme and Sloterdijk, Jaques Rancière makes no reference to atmosphere; however, his notion of the 'distribution of the sensible' was important to my own analysis of atmospheric resonance and the manner in which we make sense of the world.

3 A BERT is a means of capturing a subject's responses to a specific design. A spectrum of potential answers between two binary terms is presented to the interviewee who must select one of them.

4 Approaching people in situ is often to encroach on their leisure time, meaning that an adaptable method is essential to collecting as much quality data as possible. The BERT method enabled me to add a qualitative dimension as and when interviewees were willing, as well as encouraging self reflection from the initial answers they would provide.
5 Hyper-lapse is a technique that involves shooting time-lapse whilst the camera is in motion, resulting in a double movement.

3 Transformative participation

Sitting at the top of the newly opened Oslo Public Library, the 'Deichman' as it is referred to locally, I search its archives of local planning and architectural literature. The Deichman is the latest addition to *Bjørvika*, affording new perspectives of the area and the OOH. I gaze out through large moveable blinds and see figures parallel to me, some 150 metres away, walking the roof of the opera. Below, on the stepped seating of this living room of the city style space, an eclectic group of young individuals have spread themselves out to read, play games, or just talk. This is a modern library that encourages activity, not quiet, each floor with its own interactive theme and function. Even in these COVID-hit days, the space is well used, and it is noticeable just how diverse Oslo's population has become; groups from all ages and varied backgrounds wander around to explore their new urban space. I was told yesterday by an employee of the OOH that 'the Deichman, like the Opera, feels like a house for the people'. This is my sixth visit to Oslo. A year ago, the Deichman was only a shell of its present state whilst the first time I visited Oslo, back in the early 2000s, this area was dominated by industry and a main road that divided the city from its waterfront. Beyond the OOH looms another new structure, the new Munch Museum. Less popular with those I have spoken to, the Munch will open in the coming months. To some extent then, over the past few years, I have been witness to an ongoing process that continues to shape Oslo's waterfront, Bjørvika, and the Oslo Opera House. This process has been extensive, at times contentious, and as we will reveal, continues to live with individuals as part of their understanding of the OOH.

This chapter has a dual intention. Firstly, it aims to give the reader context by providing some historical detail about the OOH, and secondly, to explore how local people's exposure to the process of its development constitutes their attunement to the building. By unfolding how key events situate themselves in the public consciousness, I reveal a mode of participation relating to the indirect involvement of users and a wider public (Jenkins 2009, 13). This I describe as, 'transformative participation', a term borrowed from Jeremy Till (2005), although applied quite differently. As we shall see, the process of the OOH's development was a long and drawn-out affair involving collaborations between multiple players and stakeholders. The scale and complexity of the project was

DOI: 10.4324/9781032668802-3

Figure 3.1 Construction work around the OOH with the *Deichman* library behind. Photograph by the author.

such that the public's direct participation in the development process was limited. That said, I met several individuals with connections to both the OOH's development and the current institution.

Norway's model for urban development invariably involves transparency and open negotiation, explicitly following the 'political ambitions of democratisation' (Askheim et al. 2017, 587). Indeed, it has been argued that Norway's socio-economic interventions are most successful when implemented through the processes of planning (Falleth & Saglie 2011, 61). This reflects a tendency in Nordic lands where the 'political democratic system is instrumental in creating a consensual model of public space' (Nielsen 2019, 39). Consequently, Norway's citizens are typically knowledgeable about the OOH, expressing strong opinions about its value and meaning, even as they might only have visited the site on a handful of occasions. In fact, I spoke to several first-time visitors from outside of Oslo who arrived with a surprising amount of knowledge concerning the OOH. In this sense, it becomes difficult to separate people's situated experiences from their preconceived notions of the building, its ideologies, and narratives. It shows how the process of the OOH's development exposed its citizens to an ongoing discourse which has embedded the Opera building in the collective conscious, forming a general consensus of meaning. In Rancière's terms, these historical events and forms of knowledge play a key role in the distribution of a particular cut in the sensible. Thinking atmospherically, it reminds us that, whilst we are always *in* atmosphere, its boundaries can be made more, or less, porous.

Participation literature reasons that without the user's direct involvement in the process of a design, the finished work will negatively determine the user's mode of practice. In architecture, this signals an inegalitarian outcome wherein those with power simply enforce their own desires through the controlled production of building (Kaika 2011, 972). As a result, most participatory approaches posit the necessary contribution of future users to the final design, giving people the 'right to participate in the shaping of the worlds in which they act' (Roberston & Simonsen 2012, 4). This shift of power from designer to user draws upon earlier participatory literature, for example, Arnstein's ladder (2019) and its critique of the top-down decision-making process. In more recent times, 'the new participative approach to planning' pursues consensus between users and planners (Richardson & Connelly 2012, 78). These two general positions differ in that the former desires a role reversal that gives the user the power, whilst the latter streamlines the user's participatory role in pursuit of consensus. What they share is a focus on the design process as the point at which the user is handed a varying degree of agency.

In this chapter, I employ the term 'transformative participation' borrowed from architect Jeremy Till and his writing on 'the negotiation of hope' (2005). Till uses the term 'transformative' to recommend an approach that emphasises 'negotiation' as a key participatory tool in the building process (ibid., 35). He argues that in the course of architectural design, equality can never be achieved because the knowledge gap between experts and lay people is too wide (ibid., 27). To remedy this, all parties must be included in negotiation, thus balancing 'the expectations and futures of participants' through 'a potentially contested but ultimately positive process' (ibid., 35).

Here, participation is explored from a similar perspective but included is the realisation that participation begins before the decision to participate is taken (McCormack 2014, 187). In so doing, I show that explicit participatory mechanisms are not necessarily required for individuals to feel empowered, and that participation is not only the conscious invitation and decision to act but also the way the fabric of the sensible relates to an extended process of shaping what we see and hear. As participation unfolds in-situ, the resonance between dominant histories and material space becomes evident, showing how atmospheres collect and focus meaning across various temporalities. I therefore make account of a mode of participatory practice that is weak yet *transformative*, in the sense that people's expectations and futures are defined through a transparent, albeit indirect, negotiation. Focusing on the potential of a weak form of participation akin to the 'implicit participation' people feel when taking part in something that joins them to a wider context (see Gormley 2008) also reveals how participation with urban space is, 'a co-fabrication that involves more than the individual participant' (McCormack 2014, 188).

In order to better understand how this particular mode of transformative participation has played out gradually over several decades at the OOH, I trace people's understanding of some of the major events, debates, and points of contention showing how 'particular spatial-temporal distributions of sensibilities make some things visible, and seen in particular ways, and others not [...]' (Rancière 2019a, 13). The chapter unfolds in sections that outline the key historical events, actors, points of contention, and decisions made during the OOH's development. By working through these key points of interest I add historical context and background, adding a layer of user data that reflects the way historical events are active constituents of people's perceptions and experience at the OOH.

The road to Bjørvika

The OOH was completed in 2008, occupying 38,500 m^2 of reclaimed industrial harbour whilst its roof area is covered with 19,000 m^2 of marble, and 6000 m^2 of aluminium (Hammer 2011, 89). Aside from the opera stage, its 1000 rooms house over 600 employees from some 50 different vocations (Ramboll 2020).

The building sits on the water looking west at the city, underlining its role as a nodal point that brings the city together with the sea (Nytt Operahus i Oslo 2002, 8). As the building is built partly over the fjord's water, steel pilings were inserted 55 metres down into the bedrock to stabilise the structure (Ramboll 2020). As Snøhetta explained to me, this involved an extensive process of excavation and water purification as the fjord's sediment was found to contain poisonous mercury-based toxins, the leftovers from Oslo's fabric and dying production historically located up the *Akerselva* River.

For the people of Oslo, the idea of a new opera house had been machinating for decades. In fact, the idea that Norwegian opera should have its own house first arose in the 1920s but had to wait until 1958 to find its location in the rented *Folketeateret* that would be its home for the next 50 years (Christensen 2015, 739). Years of discussion concerning its location, cost, appearance, and function continued even after 1999 when the Norwegian parliament passed an act confirming the construction of a first home for Norwegian Opera and Ballet (Mortensen 2011, 7). The company had long been seeking a more appropriate platform from which to build their art and house its growing company. Their main arguments for expansion were grounded in identifiable artistic, public, and workplace needs that could only be solved by the construction of an entirely new opera building (Statsbygge 1994). Several locations were considered, eventually leaving three main contenders with the additional possibility of remaining at Folketeateret after an overhaul of its space. Of these options, Vestbanen and Bjørvika were deemed the most attractive. However, a report from 1990 that analysed the suitability of multiple sites concluded that Vestbanen was the safer bet as Bjørvika came with several unknown variables such as the quality of the fundament, and the presence of a motorway that would present enormous cost and difficulty if it had to be redirected (Statsbygge 1994). Vestbanen, located near the town hall, was the preferred site for Norway's right-leaning Conservative party, whilst politicians from the eastern districts of Oslo argued for Bjørvika (Hofseth 2008, 102). Therefore, from its very beginning, the OOH was a politically polarised object of contention, widely played out on the media stage. According to Christensen's detailed analysis of the OOH's development (2015), the process was played out against a backdrop of shifting power, inconsistent views, and multiple actors, none of which were able to force their will on the others. Norwegian newspapers presented the political divisions and debates around the two main sites of Vestbanen and Bjørvika in accordance with the differing views on the OOH's actual purpose. The conservative view, and party, supported the Opera house for mainly cultural reasons, whereas the idea of an opera as a medium and catalyst for urban development was interesting to the more progressive leftist parties (Hofseth 2008, 102).

Eventually, there was a realisation that an opera house as a modern institution would comprise a diverse workplace and potentially attract a new audience. This more expansive conception bestowed a societal duty to the OOH's role as a cultural institution, placing the project under even greater political and public scrutiny (Statsbygge 1994). From these parameters, politicians and institutional representatives collaborated under public examination to develop an Opera house very much in public view. Over time, the drive to redevelop the city saw the construction of the OOH as an opportunity to overhaul the otherwise neglected area of Bjørvika. Planners noted how 'Oslo has been socio-economically divided between east and west' and planned to make an 'axis of culture', where iconic culture-based architecture would make up 'a network of pearls', that would unite the two sides of the city (Norwayexperts 2021).

In 1999, it was therefore decided to build the new opera in Bjørvika's harbour district (Almaas 2008). This brought the OOH within the city's greater vision of 'Oslo as a Fjord City', a project underway since the early 1980s (Bjørvika 2021). The state's main consultant on building issues was Statsbygg, also confirmed as the building's client and given the responsibility of coordinating with the private sector whilst also acting as quality control. As coordinators of the project, Statsbygg was guided by political ideals that insisted 'the building shall stand as a representative institution, presenting both Norway's cultural traditions and the significance of Norwegian National Opera as a part of the nation's culture and society' (Statsbygge 2005). It was agreed that after their custody of the project, Statsbygg would hand over the finished building to the Norwegian Opera Company as the building's end user, with the expectation that they would produce around 300 shows and host 250,000 visitors every year (Almaas 2008). This relationship meant that the original client, Statsbygg, would oversee the project, eventually making way for the opera company to receive the final product, developed in collaboration with the private sector using public funds.

Transparency was key to this complex process as the architects admitted: 'there was some trepidation about spending an awful lot of tax-payers money on a building for the elite'. Part of the original justification for the OOH was that it would deviate from traditional culture projects, bringing a new and diverse audience to the institution. This approach 'endowed the project with wider social legitimacy and countered the arguments of those who claimed that opera was an exclusive, elitist cultural activity' (Christensen 2015, 741). This led to unions between politicians agreeing around the logic of investing in urban development via the Opera project (Hofseth 2008, 103) and eased concerns about giving public money to elitist cultural projects.

The Fjord City

Oslo is one of Europe's fastest growing cities with a current population of over 650,00. In the late 1980s, the city began its 'Fjord City' development, aiming to free up Oslo's extensive waterfront area from old industry and develop areas for domestic, recreational, commercial, and cultural purposes (Skrede 2016, 419). Therefore, a major part of the Fjord project involved solving questions around how best to activate people's use of a new waterfront space. In response, recreation was pursued through landscaped promenades, swimming zones, domestic housing, and business locations, opening the harbour up to multiple users. Culture was to come by way of relocating several of the city's major institutions, rehoused in iconic architecture and attracting tourists and locals alike. In and around Bjørvika, this has so far led to the new Munch Museum, the Deichman Library, and the Oslo Opera House. There was contention around the idea of gathering the city's cultural highlights at the expense of other areas. A local woman I spoke to at the OOH emphasised this point, explaining: 'I am from the area of Oslo where the old Munch gallery was, so it's kind of sad they are moving it here'. In general, she enjoyed the new development in Bjørvika but didn't believe it was founded on a 'just policy'. Nevertheless, encouraged by the success of the OOH as a catalyst for change, politicians were inspired to plan further projects (Otterbeck 2008, 21).

Whilst the OOH is generally well regarded by Norwegians, the new Munch building was often referred to negatively by the locals I spoke to. Suffering from its inevitable comparison to the OOH, individuals complained that the Munch's heavy grey tones and bounded vertical space disrupted the view of landscape, whilst Snøhetta tentatively expressed their 'surprise' at the Munch's design, pointing to their own pursuit of democratic space and horizontal architecture.

In this sense, the negative perception of the Munch Museum, which had not even opened during much of my fieldwork, reveals what people understand to be the virtues of the OOH: its accessibility, openness, and a sensitivity to its surroundings. Furthermore, the contrast between the two buildings highlights an overall tension between two main discourses of planning: democracy and efficiency. Despite a focus on participation and the local as 'a reaction to the centralized, expert-dominated planning systems of the 1970s' (Falleth & Saglie 2011, 62), there remains a parallel emphasis on efficiency, less government intervention, and 'streamlined managerial processes' (ibid., 60). This is especially evident in the first areas developed under the banner of the Fjord City, where traces of efficiency thinking can be found. Here, public authorities gave control of public spaces to private enterprises on the condition that they would uphold them, resulting in a 'tightness' of space that lacks diversity of use and users (Bjerkeset & Aspen 2017, 126). Whilst the Munch Museum is perceived to enforce the boundaries between private and public space, locals recognise

Figure 3.2 An architectural vision of future Bjørvika covers the site of its construction. Photograph by the author.

the OOH as an example of a more democratic way of building, creating an architectural icon that 'puts people first'.

Unique to waterfront developments is the opportunity to create structures with comparatively little direct association to other built forms, whilst inescapably being tied to the ecological, the meteorological, and the atmospheric. These 'blue recreational spaces' are an intrinsic part of city dwellers 'spatial experience of creative lifestyle—bathing, gardening or consuming art' (Thelle 2018, 204). As the first major structure in area, the OOH defined the role Bjørvika would play in the greater context of the city. As a guide summed up:

> it's sort of like the beginning of Oslo becoming a more modern city, modern times. A lot has happened in Oslo in the last 12 years making it more attractive as a capital, and this place has been very important in that process.

Interestingly, the guide claims that the process started with the OOH, despite the fact that the Fjord City had begun decades prior. It highlights how different buildings, under the same general project, resonate very differently with their audience. For example, Renzo Piano's Astrup Fearnley gallery (2012) situated in Oslo's *Tjuvholmen* district makes a similar attempt at combining culture with community. From the outside, some reviewers aggrandise Tjuvholmen writing: ‘this is like a little city where the visitor can be in contact with nature, take a swim, enjoy urban life, while contemplating art’ (Rosenfield 2021). However, having opened shortly after the OOH, Piano’s structure occupies an entirely different place in the minds of many locals who see its value as nothing more than a contribution to the collection of art, a building detached from their lives, or as one critic wrote, ‘the uncomfortable product of art, architecture and big business: a well-tailored envelope covering a vapid haul of corporate cultural capital’ (Wainwright 2012). Relatedly, an elder local woman told me how she experienced the OOH as ‘a bit like being in a boat’, going on to explain, ‘we have a strong connection to the sea in Norway, the sky, the heaven, it *feels* Nordic, that connection to nature’. She continued, mentioning the relationship between sea and architecture, the water lapping at the sloping edges of the building: ‘it makes me think about the water rising, if it will come up over the building. It makes me think about the environment’. Despite the more literal boat form of Piano’s structure, it is the OOH that resonates with citizens who detect notes of Norwegian heritage and a common sensibility to nature.

Bjørvika and beyond

The name Bjørvika comes from old Norse and means ‘city bay’. Located at the point where the rivers Akerselva and Alna run into the Fjord (Bjørvika 2021), its setting made it Oslo’s historic centre and main harbour, dating back to the city’s original settlements and later becoming its main transport and shipping zone (Fiskum 2003). Bjørvika’s function as a transport and industry zone transformed its topography and public access was lost as it became brownfield land[1]. Bjørvika at this time had no domestic or commercial areas whilst its motorway further dislocated the area from the city centre. The motorway itself turned out to be a main barrier to Bjørvika’s re-development (Statsbygge 1994), eventually remedied by the completion of a road tunnel that reduced Bjørvika’s traffic by 80% (Bjørvika 2021).

Snøhetta explained that a further challenge was the state of the harbour waters: ‘traditionally there has been small industry like weaving, dying, and tanneries so all the heavy metals from those materials were coming down the river and settling there’. The fjord’s muddy bottom was found to be so thick with pollutants that it required treatment to remove the poisoned substances before building could begin. In short, the Bjørvika development was always a disputed one as the different political and commercial bodies held differing visions and vested concerns. The views that eventually won out argued that locating the OOH in Bjørvika and ‘taking the fjord back’ would be an essential catalyst for Oslo’s future development, its liveability, and tourist industry. By reducing traffic and sinking the E18 motorway, space could be made to develop housing, commercial spaces, and other cultural offerings such as the Munch Museum. Thus, the political will to liberate Bjørvika won out in the realisation that it was the only means to reconnect the city with the waterfront, making it as accessible as possible for citizens and visitors alike (Vibe 2007).

Developing 'blue spaces' through the transformation of industrial waterfronts has become common practice for today's cities (See Kiib 2007). Connections to the water are especially prevalent in Nordic countries where developments are typified by a focus on public design (MacKeith 2012, 137), often summoning their industrial past (see, for example, Schwabe 2021) to indicate the move 'from industrial city to knowledge city' (Thelle 2018, 208). Bjørvika, however, appears to have overwritten its industrial past, leaving few clues of its prior state. Yet, this knowledge is retained by some Oslovians, for example, two local women who expressed to me just how much they enjoy their new accessibility: 'it gives a close relationship to the water, it wasn't here before. I have walked along the harbour for many years, and it was always three meters down, you couldn't see it'. Today, streams of people move through Bjørvika and along the edges of the fjord. Even today, aside from a growing number of boats and floating saunas, the OOH's sloping form provides one of the few points where people can move on the edge between land and water, and it is not uncommon to see people diving off its walls in the summer.

Having committed to Bjørvika as the site of construction, the discussion around the OOH moved on to the specifics of what kind of house it should be. There were fears that Oslo would become a 'mini-Manhattan' (NRK 2014), resulting in a scenario where local places become reworked into 'transnational spaces [..] that could literally be almost anywhere in the world' (Sklair 2010, 139). If Bjørvika intended to attract investment through its development, this could lead to a situation where global capital would 'squeeze out non-consumerist space and replace it with consumerist space' (ibid., 149). These concerns were expressed as the potential 'Bilbao effect' of the OOH's cutting-edge architecture. A conference was even held to debate the positive impact of a new Nordic Bilbao, set against the fears of 'a publicly sponsored eviction of low-income households from the area' (Hofseth 2008, 103). Citizens had already witnessed the promise of mixed housing in Tjuvholmen ending in an enclave of unaffordable property (Bjerkeset & Aspen 2017, 121). As a guarantee, a certain amount of affordable domestic housing was secured alongside commercial lots and the promise of a local school. This promised diversity was a key point in selling the Bjørvika vision and should therefore be seen as a deliberate deviation from older parts of the Fjord City such as Aker Brygge, which, in many people's view, had spent vast sums of money for the benefit of relatively few. This led to demands for the Opera's new home to be part of an inclusive area for all Oslo's citizens (Dagsavisen 1998)[2].

Whilst there was general optimism in a project that would revitalise 'a run-down area of Norway's capital occupied by prostitutes, drug addicts and industrial installations' (Smith & Strand 2011, 93), there was also a concern that Bjørvika might remain a 'socially strained environment', reducing the attractiveness of the opera house for pedestrians and visitors from other parts of the city (Statsbygge 1994, 52). At the same time, the vision of an opera house as 'an important symbol of what modern Norway represents as a nation' (Statsbygge 2005) underlined its social ideals and gave momentum to the project.

The current manifestation of the OOH was, in part, selected because it resolved the issue of connecting different places and demographics in the city, 'generously giving back in the form of an extended public space' (Vollsnes 2010). Within the debate around how Oslo's planning practice should proceed, and with an emphasis on deliberative democratic processes (Falleth & Saglie 2011, 60), the OOH could claim to be a symbol of how citizens wanted their city to be. This is exemplified by the way individuals would often return to mantras of the Fjord City and Bjørvika, exemplified here by Ingrid, a local woman I encountered in the foyer:

> It's like a bridge between the mainland and the Fjord opening up. So, I feel like it's a building that invites me to relate to my city. Not all choices have been good, but it used to be the biggest highway in Norway, right here. It was horrible. So, I would say this building has been part of the process that has changed the way contemporary Oslo looks. This building is very much inviting you to come. It doesn't matter if you like Opera or not.

Tellingly, for Ingrid the invitation to participate with the OOH resonates through its design but is also clearly linked to knowledge of the OOH's historical context, the ideals it communicates, as well as memories of Bjørvika as a place unfit for humans.

Snøhetta

Architect firm Snøhetta won the international competition for the OOH. Established in 1989, Snøhetta name means 'icecap' and refers to the highest mountain in the Dovrefjell range, located 240 miles outside of Oslo (Filler 2013). A majority of locals I spoke to have some knowledge of Snøhetta, for example, a local woman explained: 'they are named after a mountain, the third highest mountain which means the hood of the snow'. We came upon this topic through a conversation about her perception of nature themes in the OOH: 'ahhh, I guess it's supposed to resemble an iceberg', she said, which in turn made her think of the architects themselves.

Snøhetta, originally a small group of landscape architects, define themselves as 'a non-hierarchical collective' with 'a more democratic workplace atmosphere' than other top-down structured studios (Filler 2013). Like the OOH itself, lines can be drawn to notions of the Norwegian landscape. The company's name and image express something similar to what Norberg-Schulz describes as 'the forceful expression of the Norwegian environment of the far North' often expressed in the architecture of the south (1996, 35).

Despite Snøhetta's global appeal and success, Norwegians often identify Snøhetta as one of their own, as one visitor told me: 'We all know Snøhetta, and I think we are proud of them. I don't remember the other suggestions from the competition, but Snøhetta know the Norwegian mentality better'.

Snøhetta's original breakthrough came in 1989 when they won the contract to build the Bibliotheca Alexandrina in Egypt (Snøhetta 2021d). After the OOH was built, Snøhetta assumed a place in the local zeitgeist with its Oslo studio located just across the water from the Opera, the company is referred to with familiarity and I met several individuals who had visited the studio themselves. As Norwegians, Snøhetta is perceived to share an understanding of how things should be done, and this encompasses what the company creates and how its work is decoded by locals. This is exemplified by the reflections of one astute visitor who noted how the OOH 'has a very Nordic feel, maybe it's supposed to be an iceberg, and everything is built around snow and Nordic light. I might have answered it felt differently if the OOH was placed in another city'. Looking at the designs for the library in Alexandria, its light open space, and blonde wood interior would also look at home in a Nordic context. On this same point, a visitor reflected upon how she thought locals interpreted the design of the OOH:

> People will make it their own wherever it is, they will find things that resonate with their culture. If it wasn't so white, this could be a dry landscape, it could be desert, it could probably still be a desert with very white sand. Or like the steps of Mongolia, big open landscapes.

Snøhetta explained how the design for the OOH was intentionally minimal, devoid of anything that people might want to take home with them. Its simplicity can be interpreted in different ways, as the quote above suggests. However, for most locals, the design becomes read as grounded in an understanding of Norwegian values. Jan, a visiting Oslovian, emphasised this point by saying:

> I think if the Opera house had been in another country, it would have been closed off. In Norway we have access to nature, you can camp wherever you want as long as you are not too close to people's homes, so this is like that.

Here Jan links a feeling of openness and access to Norwegian nature to Snøhetta's architecture as an expression of democratic values.

Snøhetta are quite open about how ideals guide their designs, explaining to me that: 'axial architecture doesn't let you physicalize the journey and make it your own; the symmetry makes you say I'm going there. I can't do anything else, and that is not democratic architecture'. In this sense, Snøhetta follow a Norwegian architectural tradition, for example, Norberg-Schulz writes about

Figure 3.3 An architectural model of the OOH displayed inside Snøhetta's studio. Photograph by the author.

Åmotsdal in the north, explaining how the laying of horizontal wood beams that slope upwards from the ground results in 'a building that both rests and ascends, embodying the Norwegian relation of earth and sky. Here, forms speak the language of the land' (1996, 36). These poetic observations chime with the OOH's form, where the tension between monumental and dynamic architecture is held together by stretching the structure across the land, forming a vast horizontal that disrupts our expectations of a building. It discloses how the OOH taps into a specific architectural heritage that prioritises a building's 'genius loci' (Norberg-Schulz 1984) and the way architecture should resonate with its surroundings. In turn, these themes are acknowledged by interviewees such as Mira, who noted: 'it's Nordic because of the design, I think about snow and ice, and Snøhetta with their very Nordic style'.

Perceiving Snøhetta's 'style' to be Nordic also highlights certain generalisations which relate to material use as well as its form and function, as a young student from the USA stated: 'I can see the minimalism, the wood, they are going for that Nordic style'. International visitors often recognise the OOH as a combination of a local Nordic aesthetic and global modernist architecture. The OOH's clean geometric form plays into this perception and was highlighted by several individuals who read it as a Nordic take on an international style. However, for Norwegians, Snøhetta's Nordic credentials aid the OOH in avoiding being cast as overtly iconic, and its cutting-edge modernist association is countered by a perception of Snøhetta's human approach to building.

Citizens read its specific vernacular characteristics, constituting a notion of Nordicness associated with accessibility, transparency, democracy, and a particular sensibility to nature.

A Nolli map

On a visit to Snøhetta's Oslo office, I talked with an architect about their approach to building public space. They mentioned the importance of building on the horizontal and their belief that public space should not be, nor can be, designed vertically. It was said that vertical space disrupts the realisation of public space: 'artistic buildings, cultural buildings, should be horizontal and as accessible as possible as they are part of the public realm'. These ideals were manifested in the OOH's design, giving form through an 'accessible roof and broad, open public lobbies' (Dykers 2012, 385).

In 1748, Giovanni Battista Nolli produced his famous map of Rome, showing the open connections between inside and outside space (Sennett 2018, 218). Snøhetta use the Nolli map as a motif for the tensions between public and private space in the city. The simple figure ground technique uses black for built space and white for open space, presenting a bird's-eye evaluation of the city, its accessibility, and democratic potential.

Reading the OOH as a point on a Nolli map and looking down on Oslo as a graphic representation of social space, the opera re-takes a vast section of land for public use. Its white marble, always a tone brighter than its surroundings, literalises the designation of public space. Furthermore, it represents a potential movement from the city to the roof, underlining, 'a monumentality not constituted by the spectacular, but arising in the "force field" between architecture and its use' (Haugdal 2020, 16). Snøhetta explained, by making a foyer that is open even when performances are not taking place, and by building a roof that cannot be closed off: 'the whole site becomes part of the white again, you have given it back to your society'. It is easy to imagine how popular a café on the roof might be, but Snøhetta resisted ideas to cash in on the prime location, delivering on their ideals, and as Richard Sennett has written, opening the membrane of space, making borders porous, and allowing the in-and-out movement that keeps the space alive (2018, 220).

Some individuals recognise the fluidity of movement afforded by roof but draw a distinction with the inside: 'it's Nordic outside but inside everything is international', said one guest. Another visitor, Daniel from Argentina, recognised that, 'inside is much more sophisticated, more expensive. The nice stuff is on the inside of the building, there is no need to show it off'. Daniel was staying with Norwegian friends and had visited the OOH several times already. Talking about the building, Daniel mentioned how the architecture manifests a kind of *janteloven*[3]:'I really share that concept', he said. Janteloven was also mentioned by others such as Nora, an Oslovian, who described how 'Norwegians are traditionally people who don't brag. You should be humble, so don't show pride or social status, and very often enjoyment is communicated through interacting with nature in different ways'. Despite its iconic status, both Nora and Daniel perceived the casual openness of the OOH as a particular expression of Norwegianness.

One point of contention for most interviewees was the new Munch Museum that now fills the skyline directly behind the OOH. The Munch's height and weight were unanimously seen as an unsightly contrast to the OOH's brilliance. As one young local put it: 'the big white building next to the ugly Munch building'. Negative judgements of the Munch were often connected to its disregard for the user, as one local told me: 'the Munch doesn't feel like a building for humans'. Another simply claimed: 'everyone hates it, in the beginning I thought it was a prison. It's grey and has no shape, and I don't know, I just don't like it'.

After it opened in 2021, it was reviewed with similar derision, and critics attacked its lack of Nordic qualities, noting, 'Here, there is only glass, steel and aluminium as far as the eye can see' (Bjerk 2021). The Munch's use of materials is evaluated as something other than Nordic, a building that closes off space and overlooks the human. There is little no genius loci here, leaving a structure that feels out-of-place, the antithesis of the OOH. In contrast, perceptions around the OOH's positive democratic nature appear to have solidified over time. This was emphasised to me by Astrid, a guide who had worked with the Opera company for many years. She recalled how public opinion was originally sceptical about the huge investment in Opera that would benefit relatively few. Indeed, Haugdal relays how Norway's public saw Opera as the least deserving of Norway's institutions for major investment (2009, 216). However, the architecture has swayed people's opinions, as Astrid explained:

> People don't see it as an opera house any longer, which is exactly what Snøhetta wanted. But when we were building it, people didn't know that it would be so much more than an opera and were saying, why should we use so much money? So, I don't know what it symbolises but now it is definitely much more than a ballet and opera house for everyone.

Another guide gave a similar account, saying:

> It was a clever thing the politicians did because it was tax-payers money, we own it. Snøhetta recognised the requirements laid down by politicians and planners to make a building with clear social functions that could exist alongside the *art*.

From Astrid's optic, any debate over the OOH's investment has been entirely resolved:

> If I have visitors from outside Oslo, I always bring them here, its spectacular, architecture wise, culture wise, social wise, it's a signal building.

> And they are always impressed, and proud, very proud that we have got this, people from all over Norway, it's ours they say.

This sense of Norwegian pride relates to the success of the building and its international acclaim, but it is grounded in an understanding of specific cultural and social ideals, a human-based take on iconic architecture, and the manifestation of public space.

Competition entry 04321

In 2000, an anonymous international design competition was held for the OOH. Three hundred and seventy-nine international architects entered with 240 finalised designs (Nytt operahus i Oslo 2002, 4). The many entries attracted long queues when it was opened to public viewing (Tostrup 2009, 30). Snøhetta, a local architectural studio, was eventually declared the winner. One year later the finished designs and schematics were delivered, and by 2002 Norway's parliament had voted to carry out Snøhetta's proposals and confirmed the allocated budget. In 2003 building work had begun (Almaas 2008).

The competition's jury panel established criteria from which they would evaluate the various entries. Studies of the Bjørvika area were supplemented with visits to other Nordic countries and select European sites where significant cultural buildings were discussed in relation to their specialist design, functionality, construction, and operation (Nytt operahus i Oslo 2002).

As outlined above, the extensive process that culminated in the present OOH began with an idea of a building fit for purpose and a state-of-the-art theatre that would take Norwegian Opera to the next level. This idea became conflated over time with political ambitions that saw the opportunity for further city development and the OOH as a 'flagship institution to lever wider effects' (Smith & Strand 2011, 107). It was through this backdrop of multiple perspectives and the tensions between them that the OOH was born. With this context in mind, Snøhetta's entry was selected for its 'holistic' solution and response to the challenge of building a structure that was both iconic and open to the city (Nytt operahus i Oslo 2002).

Snøhetta relayed to me that the building was always going to be iconic: 'it's Sunday architecture, it's not weekday architecture. It's got to stand out, it needs to be monumental, but not dictatorial'. The architects wanted to seize the opportunity of making such a landmark building but wanted to do it 'on their own terms'. Essentially, the opera was conceived of as 'a social space that makes the artform as accessible as possible. But it should also act as a transitional element, taking you from an everyday experience to an artistic experience'.

To maximise the OOH's fjord location, Snøhetta proposed to build on the water at an area of Bjørvika called *Krankaia*, echoing the greater Fjord development plan of bringing the city and fjord together by selecting a site 'partly on land and partly in water' (Nytt Operahus i Oslo 2002, 8). The foyer's wooden *wave wall*, which surrounds the entrance to the opera theatre itself, marks the physical line where land meets water and the OOH hangs over the fjord (Vollsnes 2010, 348). Its design underscores the foyer as a public space, extending the narrative of sea and land to the conjoining of private and public, actualised through the building's architecture.

Snøhetta's proposal was entirely compatible with the vision stipulated by the jury members who stated that Bjørvika should be, 'rediscovered as a living place to the joy and benefit of the city's users. Access and visual contact between the city and the fjord are to be re-established'. Access was deemed part of a 'flow of movement from the city to the fjord and from the fjord to the city'. Designs that simply emphasised the building as a landmark, or that ignored the 'essential holistic policy' of the brief were dismissed, underlining the competition's desire for multifunctionality and openness. The designs that were given consideration were those proposing to expand the public space of the city, providing movement and drama through 'beautiful arrangements' (Nytt operahus i Oslo 2002, 8-10).

Today, users overwhelmingly express a feeling of openness at the OOH and how the building, water, and landscape are united. Kari, a mother walking with her friend and two children, told me that they 'always meet here. Every time we come to Oslo, it's the place you have to come. It's beautiful here, it is. Maybe it's the sea and the water'. Others also recognise the coming together of design and ideals: 'you can feel the values in the building', a Norwegian named Anne assured me, going on to mention the OOH's *folkelig* or popular feel, and the way it was 'made for the people'.

Traces of the OOH's design brief can be commonly detected in the terms used by individuals that reflect the rhetoric of the original guidelines (see Tostrup 2009). Tour guides in particular have a connection with the narratives of the architecture, and I became aware of like expressions being used to disseminate the *story* of the building. This begs us to consider whether individual interpretations of the OOH can be separated from a certain distribution of the sensible which leads to people's internalised understandings of a building that has unfolded over time, even as it continues to do so. The transparency of Nordic development projects means narratives and contentions around projects become part of the social consciousness. As one guide summed up for me, Snøhetta understood 'how to use a building as part of city planning, to promote or make a forgotten part of the city come alive again'. They succeeded by 'keeping people in mind', giving the project a 'Nordic interpretation' of international architecture trends.

The art

Norwegian building legislation stipulates that Public Art Norway (KORO) has the right to 1.5 percent of any state construction project. This gave around 30 million kroner to the commissioning of artworks, ranging from poetry,

Figure 3.4 Costume designers display their craft to passers-by. Photograph by the author.

books, film, sculpture, the marble roof, and the outdoor aluminium cladding. Some of these works, as I will explore in later chapters, hold implications for participatory practice through their interplay with architecture and the user.

Of the nine projects, most are by Nordic and Norwegian artists that work with themes relating to the architecture's connection with participation in civil society (Mortensen 2011, 10). One artwork in particular, a book titled SITE SEEING, emphasises the social aspects of architecture by documenting the OOH's emergence, creating a visual account of Bjørvika before and during its development (Schønsby 2011, 136). Ken Opprann's photographic account details the workers involved, giving them human profiles. From divers to glaziers, individual names and faces emphasise the effort involved in such a large-scale project. By pointing the camera inwards, the book makes a virtue of the inclusion of those who are usually written out of urban development projects, deleted from representations of finished architecture. Such work emphasises the OOH as a house for the people, made by people.

However, despite its intentions, such works largely disappear from public view, becoming an archived memory now hidden from sight. Other works, such as Eliasson's *The other wall*, receive extra attention due to its visual prominence in the foyer, as well as the artist's global status. In this sense, certain artworks become part of the building's ongoing narrative and experience of place, whilst others are almost invisible.

One of the more successful ways of keeping the art in the public eye is not through a commissioned work but part of the decision to place the opera's theatre productions, the craft of costume and set design, at street level. At the building's rear, the design workshops are revealed to passers-by, giving a glimpse of what would usually be hidden from the spectator, except on the stage. Workshop employees have responded by placing their objects in full view, creating a never-ending exhibition of stage art showcasing the skills involved in the production of high culture. The workshops develop a similar focus on the social aspects of architecture as Opprann's photography but allow it to be played out indefinitely and in public view.

It has been pointed out that the public dimension in Nordic architecture has been made implicit in modern times (MacKeith 2012, 140). But this has coincided with 'an inflation of "landmark" architecture, and "flagship buildings"' (Tostrup 2009, 33). In light of this tension, integrated artworks, as with architecture, tend to deliberately reveal the scaffolding of its cultural production, a trait of participatory thinking made palpable. The art commissioned at the OOH speaks well to this theme, endorsing the OOH as a culture 'house' without a strong hierarchy, giving a democratic resonance to the project.

A stone saga

In 2004, the OOH was hit by the contentious 'stone saga', concerning the choice of material that would cover the Opera's roof. The architects had previously tested multiple options for the exterior stonework and had decided on Italian marble from Carrara. Statsbygg was initially happy with the decision, but some Nordic stone suppliers complained and received support from politicians who argued that a Norwegian cultural building should of course be built using Norwegian stone (Lundevall 2008, 72). It was a case that polarised political opinion and spawned much discussion in the media. In addition, newspapers such as *Dagsavisen*[4] fuelled the debate by offering their personal opinion that the opera should not be importing foreign stone. They cited expert geologists who pointed to the inevitable decay of Italian marble in the harsh Nordic climate. Not only was Italian marble culturally out of place, but it was alleged to be inferior to local alternatives. Because of the storm kicked up by the media, Statsbygg had the original data re-tested, sending it to Oslo University. They eventually concluded that the marble was of sufficient quality and after a final evaluation of price, technical and visual quality, the contract was awarded to the Italians (ibid.).

I heard from several citizens that the Norwegian suppliers were actually happy about the decision because they would have been unable to provide such a huge quantity of stone, and furthermore, that the estimated cost would have been far higher. The politicisation of marble as a material underlines the polarisation of perspectives concerning the OOH's role as a cultural institution, who it was for, and what it should represent. According to Snøhetta, the marble was chosen for a variety of reasons, whiteness being one of them. This was confirmed by one of the roof designers who said they had also done tests that showed the granite would appear grey when wet, whereas marble retains its whiteness. Furthermore, the whiteness of marble carries a monumental status and connects the building to the home of opera, Italy. The connections to marble as a cultural material are clear, 'it's from the same quarry as Michelangelo's David', Snøhetta confirmed.

Having successfully fought for their vision of a white surface, it wasn't long before the marble was again in the public eye when the newly laid stone began to take on a yellow discolouration. One of the roof designers told me that: 'they think it was because of the chemicals. It was really a crisis, all the newspapers saying the opera stone looks like urine and it's a scandal'. Tests were conducted and it was concluded that some of the protective agents and anti-marking chemical coatings were having an adverse effect on the stone (see Gherardi et al. 2019). The reaction to the yellowing was in part due to the previous conflicts when certain politicians reacted to Snøhetta's plan to import Italian stone. The 'urine yellow' was a confirmation of the folly and inadequacy of marble as a material in the harsh Nordic climate (Marfelt 2007).

Contained within the debate around the marble and its yellowing was a debate around notions of nationalism, elitist culture, and the value of cultural investment. It re-visited the previous contentions around an opera house as a state-of-the art institution, and the opera as a catalyst for urban development. A fear of indulging elitist culture on the one hand, and a suspicion that the state was unable to deliver quality culture on the other. These were some of the themes that are still referred to today by people I met at the OOH and have become an inseparable part of the lens through which the building is received. Such was the coverage the marble received that most Norwegians I spoke to were aware of the debate. One pair of friends reflected upon the OOH as: 'unusual, because of the materials, it's so untraditional, the stones came from Italy, it's not Norwegian stone'. I asked them if they wished the stone was Norwegian, 'no', one replied, 'it's beautiful'.

As I was researching the media coverage of the OOH's use of Carrara marble, I spoke to a librarian in the Deichman. He too recalled the debate and how geologists were put on record saying that the marble was a poor choice. He admitted that he still did not understand that, if white stone was the preference, 'why hadn't they used a particular stone from the north that is as white as Italian marble but more suited to Norwegian conditions'. It appears that for some the debate lives on. However, whilst locals of a certain age remember: 'there was a lot of talk about the stone', in general, most agree that 'now, everybody loves the marble, so it works'. In any case, talking to local Norwegians about the stone saga underscores just how much the process of making the OOH has filled the public imaginary. The history and knowledge of the OOH's conception and development runs through people's situated experience and resonates as part of its architectural and atmospheric staging.

Rethinking participation: Arnstein's ladder

In this chapter, I have outlined some of the events and decisions around the OOH's conception, realised through my discussions with people in the field and a variety of historical sources. These events form a backdrop to local people's interpretations and perceptions of the OOH, particularly those exposed to the critical discussion and debates that are part of the process of Norway's system of planning (Giulianotti 2020, 236). In so doing, I have revealed how this predominantly open and transparent system shapes a particular distribution of the sensible, which in turn, results in a form of indirect yet *transformative* participation.

Prior to its opening in 2008, the idea of a new opera in Oslo had long been contested. After a mandate was finally given to the project in the late 1990s, conflicts between different interpretations of what the OOH should be and where it should be built began to arise. The direct participation of the average local citizen in this process was limited, what Arnstein's influential ladder[5] would rank as being informed, and heard but without actual agency (2019, 25). However, despite their passive state the locals I encountered expressed a high degree of knowledge with regard to the building and its process of development, which furthermore appeared to result in feelings of pride, ownership, and identity. This unsettles the use of ranking agency in a hierarchical top-down and bottom-up framework, as it suggests that projects can be successful without manufacturing the direct involvement of the user in the process of design. Moreover, if participation always involves an invitation (Beech 2008), then here the invitation to participate revolves around the idea of an emerging Oslo, a better city that places citizens at the fore. Individuals are not directly summoned to take part but are left assured by the implicit promise that they can. This challenges the essentiality of active involvement, of what is often directed at a select group of people by Arnstein's ladder model. Instead, as I have outlined above, the lack of direct individual participation can in certain cases, be replaced by a general public consensus, arrived at through a process of open and transparent discourse over a prolonged period. Ideals were not simply bestowed on the project from above but were fleshed out in unique and changing circumstances over time. This has resulted in people connecting vague democratic ideals to the 'house', grounded in an expanded notion of what an opera institution can be.

A main factor for the productive dialogue and political contestation around the OOH's process came from the fact that major players were unable to make decisive instrumental decisions independently from coalitions and other vested parties (Christensen (2015, 749). And, a prime reason for this was the core debate over how large sums of public money would be spent. Only once the OOH became cast as a focal point for social, economic, as well as cultural development was the project able to bring the differing sides together and gain the crucial support it required.

In a sense, the OOH was the product of Norway's democratic approach to urban development, which citizens generally trust as representative of their interests. On the other hand, it can be argued that the success of the OOH is in the use of 'national symbolic features while keeping the process local' (Christensen 2015, 749). Combined, the OOH and its transparent process resonates with local sensibilities, just as democratic and egalitarian ideals are decoded by citizens from the architecture itself. In Till's vision, 'the negotiation of hope' is founded on a two-way relationship between the user and the architect where the architect becomes a 'citizen expert' by listening to the 'expert-citizen' user (2005, 33). In the case of the OOH, the process was much more complex, involving multiple players. Large-scale projects are never smooth and always involve compromise; therefore, the assumption that architects simply determine urban space through their designs requires nuance. It is apparent that Snøhetta responded to the design competition brief from their privileged position and local insight, keenly aware of the discussion around the design challenges they faced. As locals themselves, it appears the role of 'citizen expert' came effortlessly.

I have argued here for an identifiable mode of 'transformative' participation. Focusing on events prior to the OOH's completion, and as I will unfold in the following chapters, the resonance between past knowledge and in-situ embodied experience is an essential consideration when thinking with atmosphere. However, it must also be true that people's situated experience always has the potential to fall into dissonance with their preconceptions. That is to say, if the design of the architecture presents a contrasting experience to people's preconceptions of the building, what it is, does, and represents, then its particular distribution of the sensible will become re-partitioned and new atmospheres emerge. One can, therefore, envisage a time in the future, when the sensible of the current 'transformative' participation has been lost, and the OOH's potential for novel individual experiences results in new configurations between the zeitgeist of the day and the experience of the buildings future material arrangement. Transformative participation reflects a consensus around the potential of participation as much as it does the actual situated engagement of the individual, and in this sense, by using the term 'transformative' I am giving name to a particular partitioning of the sensible, honed through a particular process, and over a specific duration. Consequently, transformative refers to how people's perception of the OOH is being shaped, just as any type of participation has the potential to cause a different cut in the sensible.

From this point, the following chapters analyse some of the specific modes of participation played out in the spaces of the OOH. In their different ways, they attempt to develop an account of situated experience that retains a notion of the transformative. That is to say, the knowledge, meanings, and intentions that also shape people's situated experiences, and the resonance between the staging of the OOH and the situated acts and perceptions of participatory subjects as they come together in various atmospheric assemblages.

Notes

1 The term 'brownfield' refers to an area of post-industry which leaves behind unused and often polluted zones (see, for example, Alaily-Mattar et al. 2020, 18).
2 (1998) 'Klasseseire for Oslo Øst'. Dagsavisen 23. april, p. 9.

3 Janteloven 'is a Norwegian cultural norm that instructs people to be humble. Janteloven states that one should never try to be more, try to be different, or consider oneself as more valuable than others' (Bromgard, G. et al 2015, 375).
4 (2004) 'Svarstad Harland stanser marmorkontrakt' *Dagsavisen*. 1 September, p. 20.
5 Sherry Arnstein's 'ladder' (1969) showed the empowerment of citizens on a vertical scale from 1 to 8, where low numbers indicated a lack of participation and high numbers an increasing degree of citizen power.

4 Material participation

From the city to the roof

One bright July day, I crossed the bridge from the street to the OOH to walk its busy roof and speak to those visitors who would give me their time. On warm sunny days, people are often more accommodating as they sit alone or in small groups, moving casually and soaking up the atmosphere. One interviewee named Karl explained: 'For Oslo it's unusual, very unusual. It's the architecture, the white stone, and the low horizontal'. Up to that point, I had only heard positive reactions to these qualities, but this local expressed a negative take on the OOH's quest for differentiation: 'It feels a little bit false. For me it doesn't fit in, it's more like Miami. I like the Munch building much more, it's much more Oslo'. At the same time, Karl revealed how much he enjoyed walking on the OOH's roof, only that he disliked its disregard for what he saw as the appropriate aesthetic for Oslo's local architecture. For him, the OOH was trying to be something Oslo is not, and the whiteness of its stone was key to his evaluation that it was attempting to stand out. Notably, Karl's critique assumed a position of distance, judging the building through its complete form against the city. It is perhaps telling that the informant was a photographer working at the time I interviewed him, taking images of the OOH for a local tourist company. Perceiving through the frame of his camera, he was wary of international styles being transplanted onto his local landscape.

The professional photographer's view was contrasted by a young man I spoke to on the same day. Noah was 16 years old and visiting from a western town in Norway. He had arrived at the roof with his skateboard and camera and expressed a more socially aware perspective than Karl: 'I see a lot of happiness here. I see it when I take pictures, and the more opportunity to take pictures, the more optimistic you get. I've been taking a lot today, mostly of the marble and how it fits with the sea'. Whereas Karl, the professional photographer, considered the whole building from a distance and saw a white monument out of place, Noah noted the life on the roof and the positive feeling that arises from the possibility of finding new views to capture. The professional photographer saw contrasts, representations, and tensions. But zooming in, our young photographer was swept along with the 'vibe' of the place and the ways in which people, marble, and sea came together in new and inspiring ways.

DOI: 10.4324/9781032668802-4

In the previous chapter, I highlighted how the OOH's drawn-out process of development indirectly implicated local people, transforming their mode of participation through an exposure to events which had embedding them within the OOH's process. The controversy around the building's use of marble was one such example familiar to the Norwegians I encountered. Here we return to focus on the 19,000 m^2 of Italian Carrara marble (Mortensen 2011, 89) that covers the Opera's roof as a means of exploring the role materials have in shaping people's participation with the OOH.

As a part of Oslo's Fjord City development, the OOH was required to yoke city and fjord, to 'reconnect the city to the sea' (Smith & Strand 2011, 106). In response Snøhetta devised a vast roof area, 'a "democratic" form' (ibid.) that would return to the public the space that was taken by the building's plan, providing 'a landscape and an agora, a public and civic meeting place, a social arena and a source of experience independent of the other elements related to the function of an opera house' (Hammer 2011, 88). As highlighted in the previous chapter, the genesis of these claims can be traced to the conception of the OOH, Bjørvika, and the Fjord City project, circulated through numerous sources and becoming common narratives repeated by locals and employees alike. However, how such multifunctionality is achieved and how tensions between contrasting ideals are smoothed over are rarely accounted for. In light of this, my aim here is to better understand how materiality might constitute people's participation at the OOH. This I do by investigating marble's material and immaterial qualities, its whiteness, and considering what Böhme terms *social* and *synaesthetic* characteristics (2017, 60).

Figure 4.1 Viewed from above the scale and complexity of the 'white carpet' is revealed. Photograph by the author.

Interrogating marble through its perceived material and immaterial states is in keeping with an age-old tension around architecture's material form and its social function. In the early 20th century, dominant modernist ideas were aimed at social reform (Franks 2000, 32), emphasising the importance of functionality whilst the materials used for building were of secondary concern. Le Corbusier's attack on the decorative use of materials, inspired by Adolf Loos' 1908 essay 'Ornament and Crime' was driven by the modernist notion that a new era required a new approach, wherein, the satisfaction of disinterested emotion should be separated from utilitarian need (1987, 85). Concrete became the go-to material for modernist architecture, its cheap and functional qualities replaced carved stone and with it the 'opportunity for the deposit of fantasy' (Forty 2012, 83). Clearly, the use of materials relates to a notion of experience, and for our purposes, the staging of atmosphere.

If Le Corbusier saw the exalted nature of architecture through the mathematics of universal proportion and ideal forms (1987, 86), then today's *immaterial* architecture looks to enhance affective experience by pursuing an 'architecture of process' (Bille & Sørensen 2016, 3) where meaning arrives through human interactions with the built. On the other hand, the onus on human experience allows for the commodification of our spatial interactions with the built environment (Boswijk et al. 2008, 49), implicating atmosphere in the service of capital. Thus, the process of aestheticising experience creates assemblages of architectural materials, immaterial qualities, capital, and urban development goals. Urban experience becomes part of what Gernot Böhme has termed the 'aesthetic economy', born from 'the condition in which capitalism, having in principle satisfied primary needs, must turn to desires' (2003, 78).

The OOH uses marble in a way that is intricate and calculated, yet it hides its technological expertise beneath the activity of bodies moving through its expanse. It follows Patrick Schumacher's point when he writes, 'the withdrawal of determinate meaning is best achieved not by emptying the space but by filling it with an abundance of unfamiliar, abstract articulations, as so many invitations for creative appropriation by users' (2018, 83). The OOH's space is not filled in a material sense, its design is generally perceived as minimalistic, and as I will outline, care has been taken to remove any over-abundance. Instead, the space is shot through with a feeling of potential encounter and meaning that entices visitors to explore. This invitation to participate, in relation to contemporary architecture's use of materials and surfaces, has become of key interest to academics and practitioners alike (see, for example, Löschke 2016), giving rise to a wave of architecturally orientated writing that takes the quality of architecture to be wholly dependent on the atmosphere it generates (see Borch 2014, 11). However, such work tends to focus on the haptic and sensory qualities of experience and can overlook atmosphere's potential influence on the politics of power (ibid.).

From a participatory perspective, the way agency is handed to the user in relation to architecture keeps the dynamics of power very much in view, and the dematerialisation of architecture emphasises architecture's essential role as a 'catalyst for new experiences and perceptions' that can meaningfully determine

Figure 4.2 Hammered, grooved, and cut marble. Photograph by the author.

people's spatial participation (Klingmann 2007, 2). From the designer's perspective, foregrounding experience tacitly suggests democratic agency and the opportunity for the individual to determine their spatial practice. As Snøhetta state 'people can take their own journey'. However, we should be cautious of atmospheric design as a tool for determining our experience of architecture as it may lead us to 'confuse the actual with the virtual, feelings that are hardly our own yet interpellate us nonetheless' (Foster 2013, xii). The concern here is that using materials atmospherically can 'bewitch' us (Böhme 2016, 146), immersing us in experiences that we may or may not desire. As we shall see, the marble on the roof of the OOH holds an ambiguity that arrives from atmospheric attunement to the social, symbolic, and representational. And, by focusing on the atmospheric characteristics of marble, I will now move to reveal the way these diverse and seemingly contrasting perspectives can co-exist.

Marble: surfaces of the white carpet

Italian marble is a building material with several thousand years of cultural history associated with opulence, grandeur, and ostentation (Bremner 2018, 73), making the material a perfect choice for Oslo's new opera and ballet. However, the perception of opera as an elitist past-time flies in the face of a Nordic project that shrouds itself in egalitarian values. In this sense, marble can be seen as a contentious choice, enforcing suspicions that the Norwegian state was spending tax-payer's money on cultural products to be enjoyed by the few. As stated in the previous chapter, part of the debate around the marble was grounded in a form of national identity where stone and land were attached to the people and local culture of Norway. Marble as a foreign stone was painted as a weak alternative to Norwegian granite, an ornamental but fallible material that, whilst being well suited to the Mediterranean climate, would not stand up to the harshness of the North. On the other hand, a *visible* consequence of the marble has been that the OOH is commonly perceived in terms of Nordic nature, specifically the image of the mountain. Thus, marble as a material always seems to be escaping its own boundaries, disrupting, and appeasing in equal measure, underlining the importance of materials to shaping people's situated experiences.

What was done with the marble is credited to three artists from differing backgrounds who collaborated on the project. The Opera's roof was itself commissioned as an international art project, an instance of art being fully integrated into architecture (Hammer 2011, 88). I interviewed one of the artists involved whilst walking the roof on a blustery spring day. She explained to me that the initial concept was to make a 'white carpet', making marble an easy pick. As a 'human' stone, marble appealed to the designers because 'after many years, we will see imprints in the stone here like ancient Roman staircases smoothed down by human feet'. As we walked, the artist pointed out the effect was already taking place: 'it's much more slippery here where people walk, especially by the bridge, it's been worn down. You can really feel where people have been walking'.

The artist explained to me that through a process of experimentation and inspiration, taken from local technologies of working granite, different ways of treating the marble were developed. Cutting and hammering gave the stone a unique variegated surface, catching the light in different ways. Etching into the stone was found to darken its look, allowing the designers to play with light and shadow, whilst smoothing the marble gave a reflective surface sensitive to the natural light and climatic conditions. Because of the scale of the project, they worked with large stones that could be brought together to form one unified surface. Each stone block was drawn, given a number, and then machine cut to fit together, resulting in the visual effect of a single sculpted layer draped over the form of the opera. Only up close do we notice the grey silicon seams that bind the sections of marble, a point of annoyance for the designers who desired a more natural integration[1]. The slabs of marble, each around

12 centimetres thick, presented a technical problem. To lay the slabs on the roof's rising inclines, supporting ribs had to be laid across the structure to fix the stones individually and control their huge downward force (Thorsen & Obrist 2011, 60). When art integrates so tightly with architecture, it must also adapt to, and become partly shaped by, the various technical concerns that inevitably arise.

On later visits, it was noticeable that cracks were appearing in the marble, and certain stones were being replaced, although according to Snøhetta this has been a relatively minor problem. However, it does highlight a tension between the romantic idea of a material's patina and the general upkeep of the building, as to how far materials are allowed to distort. The cracks that were appearing evoked an image of ancient marble, a desirable form of patina. In actuality, the surface must withstand, not just the soft tread of feet, but an

Figure 4.3 Some stains are quickly removed. Debris on the 'beach' after a winter gale. Photograph by the author.

array of contemporary objects like electric scooters, bikes, prams, and wheelchairs which regularly ride the public roof. On this subject the designer expressed that decay was to be encouraged, especially around its watery edges: 'We didn't want them (the OOH) to clean it, but they do. It could be like a real beach, we wanted it to be growing!' One morning, after an evening of strong wind, the edge of the OOH was littered with debris, and I was reminded of the designer's notion of a desirable patina. There was a clear discrepancy in the expression of the marble's material decay, one that reinforced its artistic and historical qualities, and the bare reality of materials I saw before me, the wood, plastic, and formica that now formed a less than desirable 'beach'. The notion of natural decay and the long duration required for a desirable patina to occur becomes a romantic designerly vision of resonance between building and outside, gesturing at a process that puts environment and user into new relations. Contrastingly, sudden short-term events like winter storms become violent dissonances that must be repaired, returning place to its original state. Thinking about how certain events are outside of the OOH's control shows that some *stains* on the white carpet are more tolerated than others.

Whiteness

On the subject of whiteness in modernist architecture, Mark Wigley writes that: 'modern architecture was never simply white. The image of white walls is a very particular fantasy. It is the mark of a certain desire' (1996, xv). The OOH's marble holds specific qualities like its historical character and a material response to human interaction, meaning an undesirable stain would be something that challenges these same characteristics, for example, the destruction of the marble's surface through graffiti or acts of vandalism.

To clarify what is meant by desire, Wigley engages with Le Corbusier's modernist text 'A coat of whitewash: the law of Ripolin' (1987) that argues for using a coat of white paint to cover the decorative. In the text, Le Corbusier states, 'art must be brought to the individual' to give them a power of judgement they presently lack (ibid., 187). So, in the layer of whitewash there is a moral act, an act of purification that gives the individual this power as 'everything is shown as it is' (ibid., 188). In this sense, whiteness resists desire, it removes the clutter of the unknown and freezes the fluctuating 'shadowy variations' of haptic experience (Ingold 2017, 103). In Le Corbusier's vision, writes Wigley, whiteness was a means of moving civilisation from the 'sensual to the intellectual' (1996, 15), enforcing a visual hygiene, 'a cleansing of the eye' by 'bracketing the sensual in favour of the visual' (ibid., 5). As Le Corbusier writes,

> If the house is all white, the outline of things stands out from it without any possibility of mistake; their volume shows clearly; their colour is distinct. The white of whitewash is absolute, everything stands out from it and is recorded absolutely.
>
> (1987, 190)

At the OOH, evidence of this notion can be found in users' social media posts where a noticeable number of objects, people, and dogs are captured against the white space of the OOH's outside areas. In these images the building's whiteness forms a backdrop to the subject, liberating it from association or context.

Consequently, Le Corbusier's whiteness opposes the kind of patina the OOH's designer spoke so enthusiastically about, preferring 'a brutally simplified form of experience' (Sennett 2018, 75). Yet, whiteness through marble, a material loaded with other characteristics is far more ambiguous, and, as we have noted, buildings are never actually white. Therefore, whiteness becomes a means of erasing certain desires whilst foregrounding others in a manner which only becomes evident when the world kicks back, as in the case of the desire for a *real* beach being shattered by the reality of a winter storm.

In the Norwegian context, modernism after WWII was never embraced as new or international but was used to create a common vision of the city that adapted harmoniously to the existing structures (Norberg-Schultz 1996, 152). In other words, Norway's version of modernism adjusted to local sensibilities. Relatedly, Hal Foster notes that examples of modernist reflexivity can often be found in the conversions of old industry into cultural and entertainment spaces (2013, IX) such as Oslo's Fjord City development. The OOH follows this blend of international and local. One of my interviewees, Nora, was a choreographer, accustomed to the physical movement of the body. In conversation, she spoke of the OOH's particular qualities as a space, explaining:

> It's very inspiring, it somehow triggers me to be creative in my body as a choreographer, I don't know exactly why. I have it also with old industrial factories, it's almost like a clean slate that invites you to physically act a certain way.

The marble's whiteness creates a continuum for people to improvise against. Especially in good weather, it is common to observe dancing, gymnastics, and other physical movements as people become inspired by the opening of space. As with the commercial products found on social media that use the OOH's outdoor area to *cleanse* the object, images of people and things placed against the marble's white expanse of space are absorbed prior to arrival. That is to say, people arrive with certain images in mind and this prior exposure to the roof's *look* informs the experiential, just as tourists seek out iconic destinations in order to capture them as souvenirs (see Sontag 1979, 9).

We have learned that marble was selected, in part, for its sustaining white colour; always whiter than the city, as well as its capacity to be worked into a 'white carpet'. And, whilst the immaterial qualities of the marble might appear somewhat opaque in the expressions of the designers, 'drawing invisible lines of force and predicting the arrival of certain intangible qualities' (Wigley 1996, 19), its atmospheric workings remain essential to understanding the way people participate with place. Therefore, it is to what Böhme terms the social and synaesthetic characteristics of materials that we now turn to.

Marble's social and synaesthetic character

The designers revealed that the roof's initial more radical design ideas were in fact toned down with the realisation that 'elegant, small discoveries would be preferable to an over-abundance of visual effects'. The way marble retains its whiteness, even when wet, made it preferable to granite which greys significantly with contact to water. Other, more immaterial factors were considered too. As well as granite being a notoriously hard material, 'its expression is also much harder', whilst marble has 'a sculptural tradition' and 'a skin like feeling', I was told that, 'marble is a soft stone that's not yet finished'. This account of marble is repeated by Snøhetta's Kjetil Thorsen who claims, 'the marble still resonates, because it's unfinished. And as it lies there, on the roof of the opera house, it is slowly changing into a denser, purer, more balanced stone' (Thorsen & Obrist in Mortensen 2011, 60). Thorsen goes on to state, 'It's not an icon. It's trying to be the opposite. Because once you allow the public to move about on the roof, it is they who generate the expression of the building, rather than the building itself' (ibid.). From an architect's perspective, the character of marble plays a key role in the human interactions taking place on the roof. More than an analogy for spatial becoming, the suggestion is that the choice of a more human material aids the production of public space, alluding towards a collaborative becoming. Building and people are moving towards an improved future, and the use of materiality, we are told, is complicit. Notably, the claims made by the artist and architect concerning the qualities of the marble seem to point in different directions, referring to both material and immaterial aspects. According to Böhme, a material's social character arrives through atmospheres, belonging to 'specific understandings of life that are cultural and changing' (2017, 62). He writes, 'Materials have a social character to the extent that they, by reason of their use of culture or tradition, stand for something' (2016, 145). Therefore, for Böhme, being aware of aesthetic sensibilities is the work of atmospheric staging as it relates to the way materials are presented to us, through an awareness of western culture (2016, 49). For marble, as we have discussed, this relates to its deep-rooted connection with classical art, architecture, and its monumental and iconic status. Historically, monuments have tended to be made from enduring materials; therefore, monumental buildings need to fashion the same feeling of permanence that can link the past with a sense of future (Miller-Lane 2000, 164). The heritage of marble, specifically marble from the Carrara quarry, was one of the reasons Snøhetta chose it: 'It's an artistic stone, I mean it's the same stone that was used for Michelangelo's David, exactly the same quarry, so you know it's connected, there is a historical, artistic, intellectual connection in that stone'. This characteristic of marble, interpreted through Böhme's writing, lends itself to 'the great festival of capitalism', a commodity material (2016, 49) that shapes an 'atmosphere that belongs to a particular form of life' (ibid., 51). Therefore, part of the work that marble does relates to our cultural understandings of luxury, high art, and splendour. Marble here is a sign indicating luxury and 'the archetype of the royal palace' (Böhme 2017, 57).

As a material used for the staging of a public space like the OOH, marble's social characteristics are brought into dialogue with its political ambitions. A guide at the OOH once relayed how Norwegian friends outside of Oslo are 'always so impressed and proud when they visit and see we have got all this. It was a clever thing the politicians did because it's our tax money you know'. Had access to the building not been *felt* by Norway's citizens, the marble's Renaissance heritage and associated splendour could have reinforced the suspicion of cultural elitism. But marble, imagined as wealth shared, apparently gives people a sense of ownership. And, in this way, the marble enhances the idea of the OOH as a social monument. As written previously, the cost of Italian marble was in fact less than Norwegian granite would have been, yet the image remains; marble is a Renaissance material with a rich universal heritage that granite cannot compare with. The architects highlight the marble, and its characteristic wealth, as potentially democratising, allowing citizens to participate in the riches of the nation previously reserved for the privileged. Marble becomes a kind of egalitarian gesture, in Snøhetta's words:

> it's saying we are giving money to you. You are walking on it. We are not using concrete because it's cheaper, it's proper stone that's going to last for 100's of years, it's yours and we are spending your money, on you.

The marble, as a luxury material, therefore, sends a social message whilst its communication of wealth designates the space as exceptional, yet because it is accessible to all the gesture is seen as a democratic one. The truth that marble was less expensive than Norwegian granite becomes obscured through marble's social characteristics and we are left to embrace another story.

Returning to the marble's monumental qualities, at some point during fieldwork I asked a pair of young Oslovians how they would describe the OOH. Unsatisfied with her companion's answer, one of them sarcastically interjected: 'ermmm you might add that it's white!'. We have already discussed how whiteness holds open the space for users, acting as a blank slate, yet discretely erases undesirable stains. Here we move to look at the role of whiteness in the constitution of the OOH as a monument. Indeed, that white is a common perception by visitors to the OOH is no surprise to Snøhetta who apparently knew the OOH would contrast with the other buildings in the city. As they themselves state, 'it will always be whiter than its surroundings'.

In Gestalt design theory, 'figure' elements are the objects of focus in a plane, whilst 'ground' provides an undifferentiated background (Lidwell et al. 2010, 96). Because lighter colours tend to move forward and dark colours retreat, from distance a white building will *pop* against its darker surrounds, making the white structure the figure and object of focus. Similarly, white architecture often has the appearance of sitting gently on the ground, serving to mark the building from its heavier surrounds (Lyons 2019, 47). This is the effect that troubled our experienced photographer Karl, as written at the top of the chapter, where his sensibility towards local architecture caused him to judge the

OOH as architecture out of place. Snøhetta intended to build an iconic structure, and clearly white marble was a part of their vision: 'you look around at your city and it is mostly greys, or slight colour tones, but not usually white' they explained. The qualities of marble allow for a simple visual effect: a means of making the OOH stand out against the backdrop of greyish Oslo and enforcing its monumental status.

Compared to the city the OOH is white. But it is also very much not white, it's yellow, it's grey, it's brown, it's even blue and in fine weather its glare makes colours disappear entirely. Yet, the overwhelming view of people I interviewed on-site was that the OOH was a distinctly white building. It appears that buildings need not actually be 'the whitest white' as long as there is a surface acting as background that contrasts all else in the world (Wigley 2020).

I have aligned the whiteness of the marble to Le Corbusier's logic of whitewash, noting the tension when we recall that a main driver of the OOH's design was claimed to be about adding agency to the user experience. Whereas Le Corbusier wanted to move society away from the sensual and towards the intellectual, the OOH is a building that explicitly attempts to re-introduce the sensual to people's experience. Rather than just cleansing the eye, whiteness here affords a space that people do not only stand in front of but move around. The materiality of the space aids the call to participate, to complete the space with the movement and bodies of visitors (Jameson in Leach 1997, 230). Whiteness backgrounds the city, inviting people to come, and when up close they find a 'canvas' to express themselves upon (Thorsen & Obrist 2011, 56). The yellowing, greying, shadowing, and reflecting that are so evident become concealed as a temporary phenomenon. Stains are overlooked, understood to be ephemeral parts of an open and transient structure that will, as the young couple I spoke to revealed, always return to a state of white. Relatedly, Wigley points out that in clearing away excess, whiteness creates 'a surveillance device scanning the very spaces that it has defined for the intrusions [...]. The white wall is at once a camera and a monitor, a sensitive surface, a sensor' (1996, xxiii). Turning the white wall horizontal implies a democratic surround, but it brings a dual capacity to cleanse as well as experience.

Snøhetta's novel use of marble appears architecturally progressive, yet perhaps in its use of materials it also hides something 'radically conservative' (Yarrow 2019, 31) that, when viewed from the outside, speaks to a lineage of modernist-style buildings. A guide who expressed an interest in architectural history had clear opinions about this: 'if you are talking about the language of the building well you could point to Le Corbusier, in many ways we wouldn't see a building like this without modernism'. He went on to identify the international trend of placing iconic cultural architecture by water to enforce its 'visual statement' but also made a distinction between these structures and the OOH:

> With international modernism you may feel that the visual idea is more important than the function, but here we have a Nordic interpretation,

> and I feel as opposed to a lot of other opera houses, I know or have visited, it keeps people in mind all the time, how people are supposed to move and react in the building or on top of the building.

From this perspective, the OOH appears to span different typologies. Malcolm Miles identifies three different theoretical ways to approach a notion of monument: 'as the imposition of an ideology', 'as landmarks or signifiers of place' and through 'democratization' (1997, 47). The question of how we can come to terms with the OOH's multiple roles is posed by Smith and Strand (2011). In their research article, they examined the intentions of government bodies and planners who saw the importance of the OOH's monumental status, a reflection of Norway's recent political take on culture which is that it should be broadened and connected to different sectors, 'in order to increase the legitimacy of public spending on large cultural projects' (Christensen 2015, 739). By including different voices in 'thematic conversations', the monument became adapted to include conceptions other than the conventional icon (Smith & Strand 2011, 102). Snøhetta explain building horizontally as founded upon ideals, but it also served as a design solution, maximising the city's access to public space whilst emphasising a move from what architecture has and does to what you feel and who you are (Klingmann 2007, 1). Highlighting the social, combined with promises of area development and future tourist investment, the OOH was always conceived of as a building with ambiguous function.

The OOH holds multiple perspectives simultaneously, and marble's material qualities are essential to this. Here, we should take into account the extraordinary way marble was used. Firstly, the designers didn't simply present the marble in its classical polished state but 'liked the idea of destroying its surface, so it doesn't become so elegant'. In other words, the classic character of the marble was agitated through a process of developing and applying new techniques and finishes that change its traditional characteristics. Secondly, as Snøhetta argue, the use of marble over such a massive area changes its perceived value, 'there is so much of it, it's the norm, and the norm point gets shifted'. The people I spoke to on the roof didn't perceive the marble as a material of decadence, which supports Snøhetta's point that the stone's sheer material excess transforms its affective qualities. In addition, the hammering treatment gives the marble an unconventional appearance, allowing it to catch natural light and create different textures. More than disrupting the sculptural qualities of the marble, hammering shapes different qualities entirely. The designer I spoke to also revealed that the decorative grooves found on marble at the highest point of the roof had an ulterior motive, added to make skateboarding impossible[2]. This was confirmed by Noah, the young photographer cited at the top of the chapter, who noted that, 'the roof feels playful, but I can't use my skateboard'. Cutting and scratching a material brings its surface to the fore and divides the material's exterior and interior (Böhme 2017, 64). The 'anti-skateboard' technology used on the marble doubles as a decorative surface that hides an intended functionality. Likewise, the hammered treatment adds utility in allowing grip for people climbing the slopes in wet conditions that would otherwise be treacherous. Interestingly, these functional qualities remain largely invisible to visitors. Placed side by side, hammered, smoothed, and grooved marble treatments encourage multiple interpretations that exist simultaneously, foregrounding one type of social whilst quietly limiting another. From afar, the detailed cuts and marble arrangements are invisible, leaving a solid sculptural form that acts as a new icon and landmark building for the Fjord City (Hofseth 2008, 101).

During fieldwork, I interviewed a guide who had been employed by the Norwegian Opera Company for many decades. She was very conscious of the way Norwegian people felt an ownership of the new house and was certain that this was the result of the roof:

> When you walk on top it's like walking on a snowy mountain, where you can see snow and ice melt down the hillside and when you are walking down the hillside in Norway, it is always like this in the nature, and the architects made that, that's why people are thinking about the ice.

The guide presents an understanding of how landscape has been staged through the material arrangement of the roof, developing an 'immaterial dimension that is of equal if not greater importance [than a material's] factuality' (Löschke 2016, 4). Speaking to the artists and architects, the deliberate creation of a mountain was never expressed, only their intent to create an open and accessible public space. Another employee noted how 'the architects have always been careful about explaining their exact intentions', preferring to let people make their own interpretations. It follows that part of designing for openness requires leaving individuals the space for their personal interpretations to come forth. We might describe this as a construed ambiguity, the outcome of an architectural approach that 'does not signify beauty but guilt' (Koolhas quoted in Foster 2013, 122). Immaterial architecture becomes a way for designers to hold meaning open, thus empowering the individual. Patrick Schumacher picks up this point, noting how vagueness and indeterminacy can be a strategy for achieving 'a soft order' in architecture, 'building on the freedom of self-directed individuals' (2018, 83). In this light, ambiguity becomes a tool for avoiding accusations of design determinacy, realising that building a space for one set of users always excludes others. Marble treated in different ways allows matter to disappear into imaginary landscapes, that as we have discussed, still resonating through its social characteristics. This is supported by noting how the aluminium facades, designed by textile designers Løvaas & Wagle, come into contact with light and weather and are sensed in similar ways.

Furthermore, the aluminium's subtle intervention, like the marble, employs a false modesty that plays down its technological features whilst 'refinement is felt everywhere' (Foster 2013, 122). In conversation with a local woman, she made a similar observation connecting the OOH with what she defined as a typical Norwegian sensibility:

> Norwegians are generally people who don't brag, you should be humble. Very few people are wearing Chanel because it's regarded as ridiculous, but you see people in the city wearing a 1000-dollar jacket that you use at Everest, so it's kind of sneaky.

The marble echoes this Janus-faced display of wealth and humility, quietly combining luxury with good intentions, just as a high proportion of Oslo's traffic appears to be made up of Tesla electric cars. The 1000-dollar climbing jacket is deemed acceptable, according to the informant, because it connects the user to the nature activities it was designed for. It has a clear function, even if it's not being utilised. What the Norwegian mentality apparently objects to is the excess of the designer product for its own sake. I suggest Marble operates in a similar manner when laid out on the rooftop. Snøhetta have aligned material splendour with a social function, without which the OOH would manifest, to re-use Böhme's term, 'the archetypal palace' (2016, 49).

As one might expect, most users perceived the artificiality of the roof's landscape, but there was some discrepancy in the way foreign and local visitors viewed the building. For example, a young woman from Poland strongly stated that 'it feels artificial, there is no nature, it's not natural, it's made of human hands and stone'. Similarly, a visiting student from the USA noted the architect's strategy 'they are going for natural, but it feels sterile'. In comparison, locals would often temper their acknowledgement of the roof's ersatz nature, citing its dynamic qualities and interplay with natural light's colour tones and shadows that change its expression.

That the roof's staged landscape is generally perceived by local users as connected to Nordic nature and not simply as a facsimile of nature tells us something about how marble's synaesthetic qualities resonate with them. Böhme writes that a material's aesthetic qualities are not found in its objective properties but in their character, which he claims is the outcome of the mode in which they are atmospherically experienced (2017, 62). In other words, the way people attune themselves atmospherically defines the atmospheric qualities of a space. This demonstrates that marble has certain qualities that present themselves but requires the attunement of those present for the material's atmospheric character to be sensed 'atmospherically'. For users familiar with phenomena like ice and snow, the roof's marble surface can elicit strong synaesthetic characteristics. Natural light, especially the low angle of Nordic light, forms recognisable structures that are attuned to by locals. A description of this phenomenological material interaction between light and snow is found in Henry Plummer's (1987, 9) poetic study of light:

> The shining whiteness of fresh snow, which our eyes simplify into pure white during the clearest hours of the day, also derives from an inner transfusion of light. The snow is composed of millions of colourless transparent crystals loosely congealed together, whose tiny facets are highly reflective yet diaphanous mirrors.

Marble itself is composed of 'metamorphic recrystallization', made up of 'extremely fine and compact crystalline structures' (Vadalá 2018, 93) that reflects light in similar ways to snow. Relatedly, Plummer highlights the synaesthetic potential of materials when he writes, 'while the material world built by man is of a different palette than nature's, it can be fashioned to similarly engage light and partake of radiant energy' (Plummer 1987, 11). Marble renders new fallen snow, splitting the colour of daylight into 'skylit shadows made pale blue and sunlit surfaces turned faint canary' (ibid., 1987, 9).

Likewise, the motif of the mountain is confirmed as corporeally felt, a sensing of the Norwegian landscape whilst on the building. The marble itself is rarely mentioned in this context, only its ecstatic qualities: whiteness, glare, and wintery northern effect. It is apparent that through the experience of being on the roof, the marble's materiality shapes a sensuous and felt presence. Such nature-driven effects are hard to define but they are recognised by the visitors I spoke to, evidencing the way marble constitutes a shift from the palace to the mountain.

Regarding the OOH as a dematerialised space, the marble becomes a medium for synaesthetic experiences to emerge, which to remind ourselves, 'are qualities that occur in more than one sensory field' (Böhme 2016, 51). A further example of how marble is synaesthetically apprehended is through people's expression of the roof's lightness and the way white stone feels light despite their knowledge to the contrary. An interviewee on the roof relayed exactly this point saying, 'I know its heavy, but it feels light'. Snøhetta utilised

Figure 4.4 The marble produces snowlike qualities in the low Nordic light. Photograph by the author.

the embodied knowledge people have of materials, explaining: 'So stone is cold, we know that. I don't even have to touch it to know it's cold, I carry that in my body'. By arranging materials, the designer can nudge bodies, encouraging them to move quickly, dawdle, and touch.

In Snøhetta's words, we find an important distinction. Their first expression entails an embodied knowledge of materials that comes with experience and prior interaction. Böhme relates this to metaphor, a material quality recognised from early life (2016, 51). Their second expression is a more vitalist account of the stone's intrinsic characteristics emerging from within. However unverifiable, this claim presents us with the idea that people can sense material qualities that go beyond the 'sequences of familiarization [from] early childhood' (Böhme 2016, 51). It is here that Böhme points us towards the idea of experiencing materials atmospherically. People sense the synaesthetic qualities of the marble through its treatment and exposure to the weather, giving rise to landscape imageries. Instead of Snøhetta's implication of a living stone emitting a type of stolid force, I suggest the marble surrounds the building in 'a swirling climate of intangible effects generated by a stationary object' (Wigley 1996, 18). The marble is inert matter, but it is not working alone; it falls into

Figure 4.5 A painting by Harald Sohlberg entitled, 'Winter Night in the Mountains' (1914). Public domain.

relations with the changing light conditions of the fjord forming a powerful atmospheric generator which becomes interpreted by people as typically Nordic.

When materials communicate through their social characteristics, as we have seen with marble, wealth, and high culture, the material remains intact as it becomes a sign or metaphor. The object can then be determined through qualities such as colour and texture, in a way that closes it off from the world, making it an identifiable object (Böhme 1993, 120). But when its synaesthetic characteristics emerge, architecture becomes 'moulded by the atmosphere rather than the other way around' (Wigley 1996, 20). Just as snow is always white yet comes in an array of colours and textures, the marble shapes an ecstatic layer that goes forth from itself (Böhme 1993, 121). Light, weather, and marble show themselves to be essential elements of experiencing the OOH's roof, just as people's participation becomes a 'partaking in the forms of life', as the environment becomes 'real' (Norberg-Schultz 1996, 51).

Whiteness and a cleansing of the eye

Earlier in this chapter I discussed the white carpet of marble and how whiteness tolerated certain patina effects only if they enforced the desirable social characteristics of the marble. I argued, following Mark Wigley, that this reveals how whiteness causes a subtle closure of meanings, just as it shapes a feeling of open and participatory space. Olafur Eliasson advises that 'when we talk about atmospheres, especially in public space, we cannot take away intentionality [that] inevitably embody an ethical stance, a suggestion about how to act and interact with others' (Borch 2014). Intentionality therefore has a bearing on the way participation is played out, even as it 'involve[s] people collaboratively creating new space and new activities in their dialogue' (McCarthy & Wright 2015, 63). Thus, from an atmospheric perspective that considers both the staging and the experience of the space, looking at the OOH's ideological posture is essential when considering the work whiteness does, what is being left in, and what is being obscured.

An image posted by the official marketing department of the Norwegian Opera and Ballet shows a Dandelion flower, growing between the cracks of the marble. The sun is shining, the sky is blue, and the sun bleaches the marble into a hazy backdrop for the yellow flower. The image is tagged with the caption: 'The dandelion has sprouted on the Opera roof; do we then dare say it is spring?' This is followed by a yellow flower and a smiling sun emoji. The image can be read as a simple positive note, announcing the end of a long winter. On the other hand, if we recall how the yellow discolouration of the marble was seen as a minor scandal, then it is notable that here the Dandelion growing between the cracks presents us with an image desirable enough to be posted by the house itself. In this sense, the image points to the tolerance of the Opera institution. Moreover, it adds an ecological sensibility[3] that tells us to slow down (see Thorsen & Obrist 2011) and see the beauty in what, in another context, would be considered a weed. Against a grass background in a park, the flower would be unremarkable but in the context of a white marble roof, the Dandelion becomes a symbol of a progressive institution. Again, the marble's whiteness forms 'conceptual skins that define potentially infinite space' (Lyons 2019, 68), inviting people to participate by using the OOH as 'a machine for looking' (Foster 2013, IX). Just as people participate by photographing themselves and other things on the roof, becoming the figure against the ground, so too does the Dandelion become highlighted to reveal its unique qualities. Atmospherically, it stages a 'coming together of materials, of intentions, the buildings; it's a hovering, a resonance' (Eliasson 2014, 93).

One comment on the image's feed responds: 'I am like that', adding a flower emoji. This ambiguous response can be interpreted as the commentator identifying themselves as an entity struggling against suppressive forces. Whatever their exact intent, it is notable that the flower on the roof becomes compared to the human individual. If we accept that the intention of the original post was to underline the OOH as an open and tolerant space for all, then this occurrence becomes coherent. Thus, the Dandelion post speaks to a 'stream of everyday life' (Jurgenson 2020, 14) that ensues beyond the elitist production of opera and ballet. It is this citizen audience that is being approached with this type of communication. As a member of the media department explained to me, whilst the goal is to promote opera and ballet, the outdoor space provides a source of never-ending images that appeal to a wider audience: 'this house, and roof, has without a doubt democratized the art form just by being so inclusive, easy to explore by foot, a hangout spot when the weather allows it, and spectacularly photogenic, even in bad weather'. If we again imagine the dandelion on the grass not on the roof, then we can see the work that whiteness does. As Wigley writes, whiteness 'is not simply inserted into the space vacated by clothing. It is itself a very particular form of clothing' (1996, xviii) which is also to say that whiteness is not the absence of design intention but is a very particular type of design.

Whiteness and blur

We have seen how whiteness forms a 'blank slate' that allows people to engage against a seemingly neutral background and how it involves the desires and intentions of its designers. As a final move, I will explore moments of unsettling glare, produced by the marble's interaction with high levels of light. A low winter sun, for example, bounces off the marble making it impossible to focus as the 'dazzling glare undoes fixity' (Bolt 2000, 208). Here, whiteness in the extreme disrupts experience in another way than we have already discussed above.

On an overcast day, the impact of the white marble is pleasantly subdued, but on a clear day, the glare can be intense. Crossing the bridge from the station side of the opera house on a sunny day can be a shock to the system. One particularly bright spring morning, I arrived at the OOH and found myself retreating to the water's edge such was my discomfort. I looked out over the fjord until my irises adapted to the situation. According to Snøhetta, a driving concept behind the OOH was to create 'a transitional element', taking people from the everyday city to a quizzical encounter with the building and its art. The scale of the project allowed for the time and space required to take visitors beyond their expectations and into a place where 'you don't understand, a space somewhere between architecture and landscape'. Light is essential to this transition as Snøhetta clarify: 'a change in light, that's why white!'

The shocking glare of white jumpstarts the senses, making the transition from the city to the OOH an embodied and felt experience. One bright and sunny day, I met Olivia visiting the roof with her son. 'You have to wear sunglasses here', she told me. This I heard repeatedly from informants familiar with the space. Light reflected from the marble gives off a harsh glare that is both all too visible and yet strangely allusive. Objects that are well lit usually become more visible to us; 'light is not perceived in itself, but through how we visually discern the varying shapes, colours and intensities' (Edensor & Hughes 2019). However, the over-abundance of light on the roof does not aid our vision, it makes things less discernible to us, just as 'one cannot look straight at the sun even though it is the source of all illumination and vision' (Bataille in Jormakka 2004, 290). Writing about the glare of the western Australian sun, Barbara Bolt reasons that 'too much light on matter sheds no light on the matter, the "glare" of the sun fractures the nexus between light, form, knowledge, and subjectivity' (2000, 204). Bolt describes her attempts at painting landscapes and finding that the glare meant forms were illegible, so fractured that nothing was revealed (ibid.). My own image-making with the pinhole camera was similarly disrupted by this condition of light. Like the iris of the human eye, the aperture of the camera doesn't cope well with intense light, reflecting and dispersing off the mineral-laden marble. Dispersion is produced in the camera

Figure 4.6 Wet marble creates an ice-like landscape. Photograph by the author.

image as it freezes the different wavelengths of light into coloured shards, breaking up the image by laying shapes and textures over the surface.

Yet, just as Bolt's writing relates to an actual landscape experience, the glare of the OOH creates a sensuous synaesthetic quality that turns the marble into an imaginary nature landscape. Instead of brightness adding optical clarity, it presents us with an overstimulation that results in, as Bolt tells us, shedding no light at all. However, this lack of clarity also relates to the experiences of snow, water, ice, and sun, especially for those Norwegians I spoke to where the harsh interaction of light and marble conjured memories and images of the Nordic landscape.

In 'the eyes of the skin' Juhani Pallasmaa writes that, 'the eye is the organ of separation and distance' (2012, 50). Pallasmaa's point is to highlight the folly in thinking of architecture as visually based when its essence is in fact haptic. If whiteness for Le Corbusier was a means of erasing the emotional and bleaching the shadows from architecture, then glare, by way of an excess of light, only serves to reintroduce such design uncertainty. Moreover, this uncertainty is essential to a feeling of intimacy that comes forth when the sharpness of our vision is constrained, inviting 'unconscious peripheral vision and tactile fantasy' (Pallasmaa 2012, 50). In other words, in certain conditions, marble produces an excess of whiteness that undoes erasure by foregrounding the material's synaesthetic characteristics, haptically sensed by those present. Böhme writes that when we 'sense' materials, we do more than touch their exterior, we sense in a very physical way that stays with us; 'the magic of the material is disclosed, even if we inevitably fall for it again and again' (2016, 147). In this way, marble moves between its social and synaesthetic characteristics. A final point, that I will also pick up in the following chapter on movement, is that this synaesthetic transformation introduces an entirely different set of social characteristics related to a local sensibility of Norwegian nature and cultural identity.

Designing ambiguity: the palace, whiteness, and a Norwegian sensibility to nature

I began this chapter by outlining two different positions from two photographers, Karl and Noah, whom I encountered during fieldwork. These photographers looked through their respective cameras and perceived the OOH in different ways. The first assumed a distanced position, an outsider's view of the building, whilst the other was present and active in the space itself. The type of images they were taking also reflected their views. Karl saw the OOH as a monument out of place, and Noah as an optimistic and life-affirming opportunity to take pictures related to the experience of place.

From this point, I explored marble as a primary building material, designed and treated in specific ways to form the extensive horizontal surfaces of the OOH. Looking at marble through its material, social, and synaesthetic qualities I have unpacked some of its ambiguity and highlighted how the coexistence of multiple simultaneous readings emerges atmospherically. More exactly, I have shown how ambiguity arrives from the marble's social and synaesthetic qualities that in turn support a democratising ideal, sharing public wealth through the luxury characteristic of marble. In addition, the marble functions synaesthetically, summoning specific nature encounters through an interplay with light and weather. Relatedly, by addressing the whiteness of marble I have explored the way the OOH's architecture affords embodied participation by presenting a 'blank slate', an open space that is not just a background but a feeling, a sensed spatial continuum that summons people to explore the space with their bodies. In turn, this encourages acts of individual representation where the expanse of white facilitates an open space for capturing novel, yet similar, images of spatial experience. Following Mark Wigley's analysis of whiteness and modernist architecture, I discuss how openings through whiteness can also result in closure, an erasure of the undesirable stains from the white carpet of the building's surface. Turning to the element of light, I revealed how whiteness can invert itself through the phenomenon of glare, as extreme light disrupts the subtle desires introduced via whiteness, obliterating them, and leaving a space open to sensorial and haptic participation. Paradoxically, glare produces the very same obscuring effect as the shadowing of architecture that Le Corbusier sought to remove via the introduction of white. Thus, it is in extreme whiteness that potential openings and emotional connections between the user and the built are found. It creates an imaginary space beyond the architecture's control, offering participatory value outside of the organisation of the OOH's tectonic space. Crucially, for locals and visitors, what fills this imaginary space is often the memories and understandings of Nordic nature and landscape, an entirely new layer of social characteristics.

Ultimately, I find the marble works in the following ways. Firstly, it becomes understood as a material with great artistic heritage and historical value, which also carries the signs of wealth and splendour. These social characteristics are used intentionally to imply the democratic act of giving back to the people, spending people's money on themselves and thereby averting accusations of cultural elitism. Secondly, the marble manifests the situated and experiential through its treatment, and the innate material qualities that interact with the natural environment in synaesthetic ways. That is to say, marble is able to shape feelings that connect users with other places, absorbing their prior experience into the situated encounter. The incorporation of the social and the synaesthetic are key to the OOH's claims of being a 'social monument'. Thirdly, whiteness is produced by the expanse of marble that binds the individual pieces into a solid form. This sculptural form shapes a monument foregrounded against the greyish Oslo skyline, but it also encourages people to become immersed in it. Here, the marble's whiteness tints the space with a specific tone permeated with the characteristics of a new social characteristic, an egalitarian version of the social. What this shows is that it is through the dynamic qualities of marble that the conceptualisation of the OOH as a social monument is wrought and sustained.

I have now made an account of how materials and surfaces constitute an atmospheric staging of the OOH's outside areas, underlining the importance of understanding the relation between designed intentions and individual experience. Investigating people's participation with architecture points to the ways spaces can be improved and made more democratic but it also highlights the potential for atmospheres to create bounded places using materials to resonate with specific cultural understandings. Following on from these observations, the next chapter will move from the OOH's material surfaces to explore what is played out upon them, by exploring movement as a mode of participation.

Notes

1 Some of the silicon seams were later sprinkled with small pebbles to hide the artificial look of the silicon binding.
2 The roof was designed before the advent of electric scooters in Oslo which , at the time of writing, had become common to the roof.
3 I will explore this ecological 'thought' further in the last analytical chapter that focuses on art as a mode of participation.

5 Movement participation

On a clear spring afternoon, I stood on the OOH's marble surface and set up my camera for a time-lapse. The day was drawing to a close and the light beginning to fade. I planned for total darkness in around an hour, timed to capture the transition between natural and artificial light. As I set up, a man arrived on an electric scooter, parked in the centre of my frame and walked away. I tilted my camera upwards towards the roof and locked its position on the tripod. Despite the cold, the roof contains more activity than I can take in. Streams of people flow past my lens, and I relax in the knowledge that their movements are now mine. Reviewing the footage, people move in waves and along certain paths. The traffic lights guarding the road from the direction of the town gather small groups before releasing them, causing regulated pulses in the number of arrivals I now see on screen. From my position, a flow of people moves from left to right and right to left. This is the path that leads from the bridge to Bjørvika, the route taken by those just passing through. Another stream makes a diagonal from the bridge to the main entrance, disappearing through one of three rotating doors. Others move from the bottom right of frame and up the north side of the roof. Along the top, heads appear over the edge as people pause for the view. High above, gulls circle the commotion against a pale cloudless sky. Between the main pathways, others deviate to peer in a window or look over a ledge. People photograph themselves and their companions. A family arrives to my right. They pose for a group portrait and move on. Two young women approach the abandoned scooter. They try to start it before giving up, retreating to sit on the nearby wall. A man passes on his bike. Several people passing close to my camera are caught looking into the lens, alerted by the click of my camera's shutter. A woman passes with a suitcase on wheels, others shoulder travel bags, a Norwegian flag flashes by. Another man stops and attempts to get life in the abandoned scooter. He also fails and moves on. A cyclist zips by. Away to the left of the entrance a man lays down his jacket on the marble slope and sits. A young man precariously walks down the ledge instead of the stairs. Two women dragging holdalls stop and rest by the wall. A lone figure all the way to my left looks over the wall to the city. A young child in a yellow jacket pushes past on a pink scooter. The warm orange glow of the wooden wave wall is starting to grow against the diminishing daylight, and in

DOI: 10.4324/9781032668802-5

the triangular window where the roof meets the ground, Eliasson's installation *The other wall* begins to show itself. The density of people eases but bodies continue to move up and down the roof. Those close to the camera appear blurred as my camera extends the shutter speed to compensate for the lack of light. Now illuminated from inside, the foyer reveals moving figures in and around the gift shop. Those outside continue climbing, jumping, pausing, accelerating, posing, and snapping. In the half-light, a young woman to my right strikes a one-legged airplane stance. Her friend, positioned above, takes her image. Together they find new poses, new ways to frame their bodies. The artificial light of the OOH becomes a kind of lantern, drawing people to it like moths. Out of the gloom, heads at the top are only just visible. A final unsuccessful attempt to use the discarded scooter is made. Now in the half-light, people outside move less, cautiously approaching a terrain without visual cues. Walkers appear as hazy blurs, only their shoes remain still long enough to be captured. A pair of white sneakers. There are fewer people around, but they have not abandoned the idea of exploring the OOH's topography. Literally, hundreds of people had entered the OOH's space in the hour that had passed. I pause the time-lapse and recall feeling the wind on my face and neck. I remember the feeling of my boots on the surface of the marble, shifting my legs to keep warm. As I leave, the scooter remains and people continue exploring the roof.

This chapter moves from materiality to focus on movement and how it invites people to participate through architecture, environment, and notions of Nordic nature. Referring specifically to the outdoor space of the OOH, art historian Erland Hammer writes:

> The roof triggers and rewards a kind of tacit knowledge, or bodily intuition, as if the artists, perhaps unwittingly, had discovered a way of marking ideological distinctions based on the mastery of nature rather than culture. It becomes a particularly Norwegian and 'good-natured' variety of elitism, based on 'universal' preferences for nature, sports, health and fresh air.
>
> (2011, 93)

Hammer's words imply a staged atmosphere that elicits visitors to move in certain ways, one that carries a Norwegian sensibility to nature that is universal yet vernacular. It is in this sense that movement, as we will go on to discuss, carries an ambiguity that supports the OOH's multiple interpretations and functionality.

Snøhetta's Kjetil Thorsen explains that the Opera House in Sydney was an inspiration to Snøhetta's thinking[1], 'before it got the big sails, it had a platform that interested us' (Mortensen 2011, 46). In pointing out that the Sydney Opera had ultimately failed by adding a sculptural form to the platform, becoming a monument rather than a public space, Thorsen reveals Snøhetta's approach to creating an architecture that allows its users to participate in the building's

Figure 5.1 Composite of a time-lapse shows the OOH as it transitions into darkness. Photograph by the author.

'expression' (ibid., 56). Snøhetta's 'horizontal' approach can therefore be seen as part of the Nordic tendency to build for 'human scale' and freedom of movement, otherwise forgotten aspects of urban development (Worpole 2000, 154). Horizontality shows an intent to open architecture up to its surrounding environment. Moreover, it breaks from the dominant gaze of the spectator, combining 'looking with walking', thus inferring greater 'democratic potential' (Foster 2013, 103). The OOH becomes a stage where the invitation to participate is communicated, and movement is essential to its continuously emerging public space.

Realising that the way people move can constitute their experience of place, questions around how and to what end bodies are coerced through atmospheric staging arise. Thinking atmospherically breaks down the distinction between individuals and their external stimulation (Philippopoulos-Mihalopoulos 2013, 36), helping us to explore how movement constitutes specific meanings and cultural understandings. In this manner, I proceed to shed light on movement's role in an ongoing process of exploration and affirmation, as people participate on, and with, the OOH.

Emblematic of the tension between building for people and building that excludes people is how 20th-century planners 'transferred the biological value of moving freely from the human to the mechanical' (Sennett 2018, 183). During the 'century of the car' (Dennis & Urry 2009), cities were shaped around the movement of automobiles at the expense of people's free movement on the street (Gehl 1987, 33). Contemporary trends in urban regeneration, as with the Bjørvika development, have challenged planners to consider movement in new ways 'to limit the negative effects of vehicle traffic and introduce innovative means to satisfy the mobility needs of the city's residents' (Marshall 2007, 41). Consequently, the so-called 'mobility turn' in academic research has focused on the social challenges brought about by these new ecologies and systems (see Sheller et al. 2019). In Nordic lands, mobility research embraces strong social and cultural interests (Freudendal-Pedersen 2007, 27), testing the balance 'between altruistic socialism and selfish capitalism' (Trägårdh 2012, 128).

Although people use a variety of transport to move through the space of the OOH, here I focus mainly on walking as the dominant method, particularly pertinent as it implies an act of leisure and non-work status (Vergunst 2016, 108). Focusing on walking follows urban theory's longstanding interest in how we relate and interact with the city (see, for example, Debord 2016; Benjamin 2015), walking's use as a critical practice, and as a means of subverting uneven power dynamics (Pinder 2011, 674). Contemporary contributors work with similar tensions around movement as a restrictive as well as liberating activity. For example, Tim Ingold writes about movement in terms of 'wayfarers' and 'transport' (2011). His distinction is based on the way humans either fall into relations with their environment or become cut off from them. Wayfarers take perambulations, embodied experiences unfolding not within places but along paths where we fall into relation with the world. In contrast, transport moves people who do not move themselves, reducing the experience of the journey to reach a destination (Ingold 2011, 148). Similarly, Rebecca Solnit uses the term 'meandering' to distinguish motility of movement from a simple locomotion between sites (2014, 3), whilst Margit Brünner (2015, 178) writes that 'roaming' 'is listening rather than intending'. Wayfaring, meandering, and roaming are all terms that describe the potential agency of movement, implicating it in notions of meaningful participation.

In Martin Heidegger's short essay, 'Building Dwelling Thinking' (1971), it is argued we have lost the innate art of dwelling to modern life. Heidegger's solution was to flee the city and build his infamous hut in the forest, a place for solitude and to escape the complexities of difference (Sennett 2018, 127). Heidegger's ideas are contentious because they developed a justification, and an association with National Socialism, for excluding others from the act of dwelling. Tim Ingold draws upon aspects of Heidegger's thinking but avoids his grievous conclusions, using the term 'inhabitation' to designate an experiential openness without limits or boundaries (2008, 1797). For Ingold, movement and inhabitation are inseparable because the world is never static, life is the unfolding of co-mingling substances and interweaving lifelines (Ingold 2007, 19). I attempt to trace some of these lines, adding a more than phenomenological perspective to develop an account of movement as a mode of participation. I begin by emphasising Snøhetta's horizontal approach to building, before exploring the way people inhabit the open areas of the roof. Moving on, I explore people's situated experiences, not simply as smooth resonances but as arhythmic encounters, shown to be essential to movement as a mode of participation. Finally, I probe how movement on the roof connects with a local sensibility to the outdoors, soft political ideals, and understandings of Norwegian culture.

Architecture of the oblique

Snøhetta explained to me how they envisage an individual's first-time encounter with the OOH: 'you cross this threshold, and you end up on this peculiarly white empty landscape which folds up and folds down to the water connecting you to the horizon and the sky'. Visitors enter and are invited to make a choice, do they go up, around, or inside. Snøhetta designed these 'soft thresholds' to physically, associatively, and emotionally remove visitors from the city, its noise, and tempo. It introduces the idea of slowing down, a dominant theme behind all the architecture and art projects at the OOH, presenting visitors with the opportunity to find new rhythms and new ways of moving (Thorsen & Obrist 2011).

People mostly arrive at the OOH on foot, but they also bike, run, scooter, walk dogs, dance, and use their body in various other ways. Some breeze past on their way to elsewhere, some meander and explore at leisure whilst others arrive to work, meet a friend, or to see a show. Those with a ticket appear to have their movement limited to finding their seat, whilst those without move in a different way. They explore the space, its views of the city, and the exposure to the wind and sky for as long as they desire. Generally, the informants I spoke to have done all of the above. Multiple visits relate to multiple reasons for visiting that accumulate and merge over time.

We are familiar by now with Snøhetta's horizontal approach to architecture that combines park and plaza, the two most common sites of democratic urban space in modern cities (Nielsen 2019, 41). Iconic buildings, like the Guggenheim in Bilbao, gesture their interest in public space, yet look inwards to '"protect" themselves from the public realm by blocking access to the public, or by making access excessively expensive' (Kaika & Thielen 2006, 63). Such structures only make sense at a distance, circulating the building as a 'logo' for the city they are set in (Foster 2003, 38, 29).

A contemporary example of applied verticality can be found in the OOH's neighbouring building, the towering Munch Gallery (2021). I visited the Munch shortly after its opening and took a series of dawdling escalators to its summit. Whilst the views are impressive, placed behind glass one is cut off from the sensation of the environment outside. The possibility of movement on one's own volition is removed, whilst the opportunity to 'slow down' is enforced by the programmed tempo of the architecture. In other words, there is little agency for the individual and the experience becomes akin to a ticketed ride, ending in 'a vertigo of disconnection', as we lose touch with the ground (Wylie 2002, 451). Apparently, inhabitation has little to do with earthly contact and everything to do with the relations one is able to sense.

In the example of the OOH's product logo, we find a black oblique line running from left to right. This simple line repeats the form of the building as it slopes from the air to the sea, echoing its connection to the surrounding environment and invoking the movement visitors make as they navigate the OOH's roof. The success of its 'proportional density'[2] is in the way it holds a

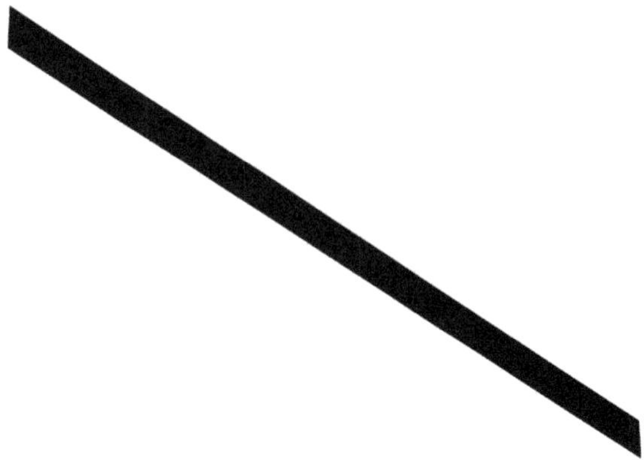

Figure 5.2 A rendering of the OOH's logo. Produced by the author.

narrative that resonates with the perceiver. The perceiver is invited to engage because parts of the story are missing that must be filled. In this sense, the logo's design follows Snøhetta's stated desire to engage visitors by affording them the chance to complete the architecture. Furthermore, the simple line can manifest a sensuous response that draws upon memories of individual movement experience. Embodied experience becomes implied through the image, just as the OOH's place between water and sky is established as a dynamic membrane, co-constituted through the potential entanglement of elements on the roof.

It's helpful at this point to consider another architectural endeavour to enhance people's participation through movement, as it reveals something essential about the way movement is shaped at the OOH. Here I am thinking about Claude Parent and Paul Virilio's oblique architecture of the late 1960s. For Parent and Virilio, their approach was a push against the wrongheaded conventions of modernist building. Le Corbusier, for example, believed 'that to travel in straight lines is to reveal the ability to reason and a sense of purpose peculiar to homo sapiens' (Banham 1960, 248). In rebuke, Parent and Virilio employed 'disequilibrium and motive instability', shaping inclined slopes that used gravity 'as a motor' to generate a 'habitable circulation' of movement to take place (1996, 12).

Parent and Virilio believed the general public's lack of interest in architecture had led to failed attempts to 'shock' people from their apathy (1996, 67). Their solution was for an architecture that would enhance bodily sensation, open communication, and engage people by drawing them into an environment, making them feel part of a universal architecture (ibid.).

Three steps were required: make people receptive to the space, encourage them to participate, and give them a sense of belonging. By challenging visitors, a 'potentialism' was revealed, moving them 'beyond refusal'. This formed a catalyst between individuals and architecture who, 'thus liberated, will be able to evolve unconventional means of communicating with and consciously PARTICIPATING in architecture' (Parent & Virilio 1996, 67). Notably, from the architect's use of upper-case, there is an emphasis on participation and architecture's role in instigating people's movements, cajoling them into bodily relations with a building that results in a form of liberation for the user.

Snøhetta express a less directly deterministic approach, although Kjetil Thorsen uses similar terms when describing the marble roof as 'heavy, and so archaic, as you move up and across it, you are in effect moving against the gravity' (Thorsen & Obrist 2011, 60). Thorsen's assertion links the characteristics of marble with the laws of physics in unsubstantiated ways that nevertheless reveal something of Snøhetta's intentions. For example, Thorsen refers to an embodied interaction, where, as Pallasmaa has also observed, 'gravity is measured by the bottom of the foot, we trace the density and texture of the ground through our soles' (2012, 62). Gravity as an invisible yet felt force speaks to movement in an atmospheric register. The roof becomes a strange synthesis of situated experience, materials, and cognitive interpretation, as movement aligns bodily sensations with specific cultural knowledge, thus blurring the distinction between bodily knowledge and cultural forces.

Returning to an architecture of the oblique, people arrive at the OOH from the direction of the station commons[3]. Crossing the narrow bridge, individuals enter the ground level of the large roof area. Depending on the weather conditions, the change in light can be dramatic, causing people to pause. New arrivals give themselves away by hesitating as they stop to assess their situation and plan their next move. In busy periods, the upper platform is quickly identifiable from below as a desirable destination point, revealed by the streams of people moving up and down its sides, but on quieter days the route is less defined, and the space more open to interpretation. This moment of pausation is potent, as Heidegger hypothesised, 'I am never here only, as this encapsulated body' (1971, 6). When we move towards a goal, we are in a sense already there, we pre-envisage our future position. This alters movement from a progression of a to b, to a movement that projects into the future, akin to the contrast between transport and the wayfarer. It opens up to potential experience and the act of physicalising a journey, as Snøhetta put it: 'I have chosen to do something and it's my choice'. However, the notion of 'choice' becomes problematic in light of the intent to implement transitional points engineered by material arrangement. The narrow bridge, sudden changes in light, and the opening of space that points arrivals away from the OOH's entrance and outwards towards the fjord conceals the actuality of two main paths, shrouded amidst an atmosphere of felt possibility.

In addition, Snøhetta play with human scale: 'making the building feel smaller than it actually is by the trick of tilting the roof at a slanting angle, obscuring the scale of the building as you approach it'. In so doing, Snøhetta pursue a modest monumentality through horizontal form, creating a less ceremonial entrance that avoids reflecting uneven power structures (Thorsen & Obrist 2011, 55). Snøhetta's awareness of the 'obscene dictatorial' tendencies of monumental buildings has caused them to re-think conventions such as doorways as symbols of power (see Tuan 2014, 42). Instead, understated portals and ambiguous paths speak to its democratic intent.

We can now begin to understand something of how the oblique horizontality of the OOH choreographs people's movement atmospherically. Noticeable too is the unusual lack of signage used to direct people's paths. 'There's a lot of things we actively removed', Snøhetta told me: 'in order to clarify and clean up the experience'. Another of the designers recalled: 'we wanted the feeling of being on top of 30,000 square meters of stone', so visual markers would distract from the illusion. Even at the top there is no fence. Safety comes through the use of a 'ha-ha'[4], a landscaping technique developed by gardens and zoos to provide an invisible barrier (Solnit 2014, 88) that blocks off potentially dangerous edges with a trench that allows unobstructed visibility. In addition, the roof angles in such a way that surrounding buildings are mostly hidden, leaving its form to meet the sky, evoking dramatic landscapes. People move around the roof seeking new perspectives, that in turn, shift their focus from the building itself, to the surrounding environment, a means of examining the world. As Hammer writes, 'the roof becomes a way of framing our gaze as much as it demands that gaze for itself' (2011, 88). This often results in people photographing at specific points of the building, framing their companions in repeatedly similar compositions. In this sense, the building presents a series of staged surprises, propelling people through the space with the promise of unique discovery.

Figure 5.3 Bodily participation at the OOH. Photograph by the author.

As we move higher up the roof, the architecture enhances feelings of freedom and openness, as one young local expressed: 'you are free because it's high up and you can walk around and talk. It's really fun when you walk, you go that way and this way'. This informant conveys a feeling of liberation from the city and the agency to choose their own path.

According to Franck and Stevens, people tend to react playfully to liminal spaces, which they define as 'an interface between two quite different spatial, perceptual and social realms' (2007, 14). The roof's design encourages a sense of liminality by clearing the space of any tell-tale signs of the city that would disrupt the fantasy of landscape. Even the wastebins are white and unmarked whilst the security chains that signal where people should not walk, are light silver and thin enough to disappear into the roof's overall gestalt. Asked about this, one of the roof designers laughingly exclaimed: 'we didn't want them, but they don't do much as you can see', pointing to several people in the act of walking through them.

The invisibility of its security infrastructure gives the roof an unregulated feeling which arouses 'the passions' (Kracauer 1995, 68), appealing to the idea of the journey which Snøhetta understood: 'people buy into as long as a certain number of things are done'. In realising that, 'a path is a prior interpretation of the best way to traverse a landscape, and to follow a route is to accept an interpretation' (Solnit 2014, 68), the architects signal their intent to open rather than predefine, stating: 'the way you build that journey is your series of choices and the journey is half of the thing you take home with you'. As people set off on their 'journey', their tempo visibly changes, and their participation takes on an exploratory nature. Jane Bennet writes that there is a degree of agency tied to the 'movement away from somewhere even if the toward-which it moves is obscure or even absent' (2010, 32). Sensing opportunity, people shift from being mobile to motile, they roam the space, no longer following predetermined lines of movement. Their tempo changes to take note of the 'surprises', discovering at an individual level, in different ways and degrees (see Brünner 2015, 178). In this new participatory mode, roof roamers can be seen to move off their *beeline*, zigzagging to inspect things that have caught their attention. There is agency in being able to make it up as you go.

From below, people look towards the roof's upper platform, transcending their current position and projecting themselves towards their future destination. This atmosphere of futurity can be interpreted as a feeling of 'possibility that resides in not knowing exactly what will happen next, remaining open to attuning to the circumstances in which we find ourselves as we make our way through the world' (Sumartojo & Pink 2018a, 98). It is a sensed possibility, essential to and generated by movement on the roof as it drives people forward. Moreover, the feeling of possibility is universal in different types of visitors, suggesting that when we navigate space sensorially we transcend cultural understandings. The architecture provides a framework from which individuals must interpret their experience in relation to the present moment (Norberg-Schultz 1996, 79). Importantly, these embodied experiences are key to people's attachment to place, as we will discuss in the coming section, and where feelings of inhabitation can manifest.

Writing about a tour through a Danish snowscape, Tor Nørretranders describes: "as we walk we slowly hone our ability to see the shades of colour and the drama of our surroundings [...] I don't invent the differences, but I spot them, I develop my ability to see them (Nørretranders & Eliasson 2015, 176)." Similarly, the surfaces of the roof and its interplay with light, manifest an environment where people slow up to experience a world away from the paralysis of urban overstimulation.

Above, I have outlined some of the intentions and material arrangements of the OOH that pertain to movement participation. How, as people arrive from the city, the 'leap' from one atmosphere to another is tempered by smooth transitions between the otherwise 'discontinuous' elements (Albertsen 2019, 11) of Oslo. As individuals enter the space, they are met with an engineered change in rhythm that ends in a sense of futurity, propelling people into the space.

Dwelling, tripping, inhabiting

The previous section considered movement through the designed intentions of the architects and how people's bodily interactions with the OOH shape types of participation. I wish to push this further by exploring notions of dwelling and inhabitation, which is to say how people become attuned to an environment in embodied and meaningful ways. I engage the concept of dwelling before zooming in to look at the arhythmic encounters of people's experience, that 'marks the open way of life more than coherence, [...] it is a dissonance for which people take ownership' (Sennett 2006, 1). This I do by exploring some of the 'mishaps' (Vergunst 2016) that occur on the roof of the OOH, seeing how inhabitation does not only consist of positive attunement but also unforeseeable ruptures. These mishaps are part of the journey that can 'generate a different intimacy, and interrelation' with the world (Lorimer 2012, 84), exceeding the direction given by atmospheric staging.

Norwegian visitors to the roof typically express their perception of a Nordic landscape, for example, a mother and her two young boys explained:

> It feels like a real space, it's like an iceberg. You get the same feeling as you get when you walk on top of a Norwegian mountain in the winter in the snow, so it feels familiar. Yeah, it feels the same as a snow-trip when you walk up there.

The quote underlines a familiarity with the Norwegian landscape, summoned through the roof's felt qualities. Importantly, Norwegians don't identify a generic mountain but identify a Norwegianness in the OOH's landscape. Drawing upon their own experiences, their apprehension of the roof is essentially indecipherable from real experience. On this claim, Ingold writes out from personal experiences on a beach, 'the waves were their sound, the wind was its feel, the sky was its light' (2011, 134), the world is reduced to an atmospheric resonance and things are as they feel. Similarly, Pallasmaa writes on the ability of architectures to produce embodied experiences akin to natural encounters: 'standing barefoot on a smooth glacial rock by the sea at sunset and sensing the warmth of the sun-heated stone through one's soles, is an extraordinary healing experience, making one part of the external cycle of nature. One senses the slow breathing of the earth' (2012, 62). Both Ingold and Pallasmaa describe affective experience through an attunement to the real world, but by reducing the very ontology of things to their atmospheric character, they assert the potential for staged environments to be equally sensuous, equally real.

In 'Building Dwelling Thinking' (1971), Heidegger 'understands the inhabitation of space and place as unfolding, emergent life, that includes logical thought and knowledge' (Jones 2009, 267). When Heidegger writes, 'only if we are capable of dwelling, only then can we build' (1993, 8), he is proposing an architecture that requires an understanding of what it means to build an opening, 'a bridge' that gathers the world and gives form to dwelling (Leach 1997, 102). However, as outlined above, Heidegger's version of dwelling was shot through with proprietorship, a clearing with borders, an enclosed and localised space, whereas Ingold proposes 'an open world, [where] the creeping entanglements of life will always and inevitably triumph over our attempts to box them in' (2008 1809). Movement is essential to Ingold's concept of inhabitation because it is the engine that drives emergence. As Solnit writes, when walking 'one lives in the whole world rather than in interiors built up against it' (2014, 9). When individuals move on the roof their encounter is not simply an embodied encounter but, 'a co-mingling where we breathe in the wind and air just as it continually tangles with the earth and sky' (Ingold 2011, 115). Movement is not ours alone, but a world in motion. By emphasising the intimate, more than sensory engagements people have with the world, Ingold leads us to a 'hybrid' landscape made up of human and nonhuman relations, dissolving the distinction between cities, countryside, and wild nature (Jones 2009, 269). In this way, the roof of the OOH becomes perceived as Nordic plateaus, an open world that emerges around the individual.

Visitors to the roof become entangled as they move, constituting and re-constituting the OOH's roof. Their movements advance partly in response to these other bodies, meaning movement is embedded in and forming of atmosphere (see Brünner 2015, 178). Recalling several windswept outings I had on the roof, the sound of water could be heard breaking against the OOH's *rocks*, whilst gusts whistled against the building's structure, providing a sensuous example of being part of a swirling, open outdoors. These moving elements combine to immerse people and shape atmospheric experience, which in

Ingold's terms is to say they no longer simply perceive the OOH but perceive with it (2011, 138) in an act of inhabitation through participation with the world. Oslo's elemental forces of water, air, and sky surround the OOH and are typically present in many of my informants' perceptions. Individuals reflect on the OOH's sensuous artifice, summoning their previous nature experiences: 'It's like going hiking on a mountain, it's like a real mountain. It gives me almost the same feeling, the view over water, but the feeling is real, I think'. Vannini and Vannini touch upon this ineffable feeling of the real, researching 'wild atmospheres' as 'felt in moments that exceed compartmentalized notions of affect, emotion, and sensation' (2020, 3). These are vital moments, they argue, found anywhere, 'in a city sidewalk, and even inside our bodies' (ibid.). The connections people make to place show how movement contributes to a sense of spatial inhabitation, that can arrive via spatial design. In this sense, designed atmospheres always hold the potential to be more than their intentions and bodies can 'move along paths yet still dwell' (Platt et al. 2021, 108). With this in mind, I turn to explore the potential for 'mishaps' to occur along the journey, with the contention that participation through movement can be better understood by highlighting examples of experiences that include dissonance as well as resonance.

Walking at length on a consistent horizontal the body can adjust to its task and relax, but on the steep climb up the OOH, walkers are left out of breath as the change in gradient takes their bodies by surprise. I experienced this myself, talking to interviewees whilst walking and being left with a deficit of oxygen as the gentle flat turned into a long slope. As people turn the corner towards the main roof platform, their quickened speech and sharp inhalations reveal the work their legs have done. The point here is that whilst the strain on the legs and lungs might disrupt our walking flow, it's 'a primary experience of nature too' (Solnit 2014, 257). Walking the OOH is not one of steady rhythmic movement, but involves changes in tempo, changing materials underfoot, and environmental conditions that the body reacts to along the way. Larsen (2021) writes of the marathon runner being subjected to a constant arhythmic disturbance from the strains of the body and the volatility of the weather. Walking the OOH's roof is evidently not comparable to running 42 km, yet Larsen's study highlights how breaks in a journey's flow can deepen the relationship between mover and place. That is to say, even uncomfortable and undesired sensations are important markers of people's participation with the roof.

When negotiating what Franck and Stevens refer to as 'loose space', people are confronted with a 'variety and unpredictability of movement, [that] requires careful attention' (2007, 14). Accidents have occurred, a roof designer confided to me: 'some people have fallen but it's mostly people taking photos and tripping'. I too observed several people tripping over the subtle changes in gradient as people focused on something other than their feet. Such incidents lend a degree of authenticity to the act of walking up the 'mountain'. The hammered marble, used on parts of the roof, gives traction as people lean into the slope, making it possible to climb safely under wet conditions. However, the slippery expression of the marble observably inhibits the way some people move. Landscape designers know 'smooth and jointless surfaces of paths can promote fast-rolling movement whilst rough, irregular surfaces with multiple joints encourage slow and careful walking' (Weidinger 2018, 28). Here, wet conditions cause people to test the ground before deciding how they should proceed, the wet marble presenting an atypical urban surface made of both smooth and irregular parts.

Hammer points out, 'walking or running through the woods is different from moving over a flat surface', implying that the OOH is more woods than pavement. Hammer goes on, 'you have to watch the ground almost all the time; you have to pay attention to everything around you. You have to move your feet more quickly, with shorter steps' (2011, 93). This describes well the way people shuffle around parts of the roof as they become aware of potentially hazardous surfaces. The observation of different material textures summons embodied reactions into the act of walking as individuals tacitly recognise the task at hand. When navigating 'a textural environment' (Vergunst 2016, 115), people slow down, they take care; being attentive to your own safety involves a degree of concentration often lacking when walking on even flat ground.

One spring morning I walked the roof with Daniel, a visitor from Argentina. He mentioned how in order to really appreciate the architecture you needed to be outside it because walking on the roof meant he must concentrate on his movements: 'I must look for surprises', he said. Confronted by subtle variations and different textures, the roamer must adapt to the situation and find the new rhythms that allow them 'to be careful' (Vergunst 2016, 117). According to Hammer, this knowledge is culturally specific:

> In winter when people walk on the roof it is easy to spot who is Norwegian and who is from a less winter-savvy country. Norwegians walk cautiously and with control, tending instinctively to walk alongside the walls when the sloping surface is icy. [...] Normally architecture does not invite us to use it in ways that make it possible to discover this kind of distinction.
>
> (2011, 933)

I cannot clarify distinctions between how tourists and Norwegians move on the roof, but the claim that the OOH allows people to participate according to their previous experience and knowledge is supported when observing the different ways people advance upon the roof's terrain. The moving body partakes in a unique journey because it draws upon 'the small skills involved in walking, relying on a direct feel for the environment', where 'no two steps are precisely the same' (Vergunst 2016, 113).

Jo Lee Vergunst has written about the 'mishaps' that occur when walking as intense moments which ultimately define the journey through the sensuous nature of stepping (2016, 108). Similarly, Lorimer describes running as a sequence of 'earthly landings' where every footfall explores the relationship between bodily force, weight, and the nature of the surface beneath, as 'the means for locomotion, pushing onwards' (2012, 83). When walking up steep, potentially dangerous inclines, each step becomes a felt connection with the ground, connecting us to the environment through our bodies.

As written, tripping is not uncommon on the roof. In particular, the slope on the city side has subtle variations in height. People generally trip up not down, stumbling but not usually falling. Most of the sloped area is of a hammered marble, giving traction to people's steps. Its sticky quality can be felt reassuringly through one's shoes, giving the confidence to stride forward instead of shuffling along exploring with the feet. Because of its irregular surface, the body senses and transmits information through its nerves, muscles, and joints (Lorimer 2012, 86). The process is an unconscious one; each individual finds their own way to navigate the terrain, haptically exploring the OOH's architecture, testing the dangers of a perceived slippery, insecure ground.

Nonetheless, mishaps are not reliant on terrain or weather. A personal experience with a large seagull that stealthily swooped over my shoulder, grabbing a muesli bar from my hand, left me momentarily dumbfounded. Spinning around looking for confirmation of what had just happened, I looked up and saw several potential culprits wheeling away in the sky above, one of which had touched my hand, leaving the scrape of its beak. My eyes met with blank stares from other walkers giving me the realisation that this was an experience I shared only with my avian raider. I had begun to inhabit the OOH's roof through my journey, having met the world in an unexpected way. In fact, my encounter with a thieving Norwegian gull remains a stand-out memory of being on the OOH's roof, showing the force of 'mishaps' in defining our spatial experiences. Furthermore, its recollection is triggered by the many images people post of the birds from the top platform, or even by the sight of gulls in other similar urban seaside locations. The brush of a ghostly entity on my shoulder, and the touch of a beak.

Considering Ingold's image of human participation in a dynamic world, extending ourselves along relational lines, I have found dissonant as well as resonant forms of attunement, revealing how arrhythmic 'mishaps' encountered along the journey are essential to experiencing a 'textured world' (2008, 1799). However, through Ingold's interpretation of atmosphere, the expression of a culturally defined nature becomes opaque amidst the swirling elements of an always emerging roof. That experience becomes innate and universalised is not a complete picture, overlooking how individual encounters are also coloured by more collective perspectives. This I will explore in the final section of this chapter, where, in light of the potential for diverse participation that atmosphere implies, I will investigate how potential difference tends towards homogeneity, explored through the cultural and historical resonances connected to movement and Norwegian identity.

Life in the Norwegian open air

As people make the ascent up the slope, it is common for them to pause before moving on. Met with the sudden 'slope and height of the roof that erases all signs of the horizon and gravity; there is only one view toward the sky' (Mackeith 2008, 87).

I have previously emphasised contemporary architecture's tendency to leave hints of the local in order to ground the building in a dialogue between structure and place, what Hal Foster calls a 'souvenir of the old culture' (2013, 66). Here, souvenirs take the form of nature landscapes and a connection with the fjord, a Nordic approach to building which attunes visitors via Genius Loci (see Norberg-Schulz 1984). This is important to bear in mind as we explore the role of culture in forming particular ways of knowing the world. This particular section of sky, pictured in Figure 5.4, exemplifies the way form and materiality renders a landscape that speaks to visitors in specific ways. These images further enforce relations between the environment and the built, constituted by the individual's exploration of these designed vistas. Views of the landscape become stitched together as informants make connections between the OOH and Norwegian landscape. In this sense, 'the roof project truly makes sense only insofar as we walk on it' (Hammer 2011, 99). Supporting this, a guide explained: 'when you walk on top it's like walking on the snow mountain, where you can see the snow and ice melt, when you are walking down the

Figure 5.4 People making their own *journey*: visitors migrate towards the sunset at the top of the OOH's roof. Photograph by the author.

hillside in Norway, it's like this in the nature, the architects made that, that's why people are thinking about the ice'.

The resonance with Norwegian landscape, apparent in local people's perceptions, therefore, summons prior individual experiences, a universalism of nature encounter, aligned with culturally embedded motifs. One dominant motif is the romantic notion of the north. As Norberg-Schulz writes, the architecture of southern Norway has long been influenced by the 'forceful expression of the Norwegian environment of the far North, it is northern Norway that reveals its deepest meaning' (1996, 35). Therefore, we can characterise the local sensibility to nature as a national sensibility, tied to notions of identity. We might understand this better through Böhme's use of 'physiognomies' (2017) and the expression of general character through surface appearance. Norway's rugged north is identified by locals who constitute a kind of 'atmospheric sociality' (Albertsen 2019, 6), their relations organised through an understanding of the encompassing physiognomies of architecture.

Hammer contends the roof constitutes a benign and open form, governing movement in a way that has always been at 'the core of the Norwegian value system' (2011, 91). Snøhetta are quite clear about their intentions to create in tune with 'Nordic think' and how it allows designs to communicate with locals in certain ways, whilst the architect's reluctance to commit to any strict reading of their work suggests the usefulness of ambiguity for architecture. The OOH can be argued democratic in the sense that it opens situated experience up to all, yet it speaks profoundly to Norwegian sensibilities in ways that interviewees reflect upon as fundamental to national values and identity.

Jørgensen cites Norway's independence from Denmark in 1814 as the point where 'landscape, culture, and national identity bled into each other to create a new idea of what it meant to be Norwegian' (2014, 275). Moreover, institutions like Norway's organised physical education programme, which followed German and Swedish models of collective gymnastics (Seippel & Skille 2019, 109), became challenged by national characters such as polar explorer Fridtjof Nansen. Nansen believed 'the heart of the nation was to be found not in the urban capital but among the people of the wilder regions of the Norwegian landscape' (Ween & Abram 2012). As an active promoter of self-reliance, Nansen insisted that individual nature experiences would form the ideal individual, unlike the choreographed and collective physical education of gymnastics, 'guiding people back to nature would test minds and limbs […] filled with the rustling of the spruce forests and the highlands' (Augestad 2020, 40). Something of this independence of thought and agency was expressed to me by an elderly Oslovian. Noting the OOH's tolerance to visitor's behaviour, she speculated as to whether other opera houses would ever tolerate such conduct, concluding they would not. She explained her general perception of the OOH as 'a bit Norwegian. For us rules are not so interesting. We have a wildness! When we cross the road, we look for the cars and not the red lights. Is a car coming, no? ok we go'. There is something of Nansen's pragmatic individualism expressed in these words.

Another common expression of self-understanding is that Norwegians 'are born with skis on their feet'[5]. Anders, a local man said exactly that to me during a conversation on the roof. As with the general Norwegian sensibility to nature, the Norwegian for skiing, *skiidrett*, connects the physical act of skiing with broader principles of national identity (Denning 2015). In this way, movement on the OOH is 'about doing and redoing memory' (see Bergman 2010, 77), where Nordic identity is reproduced in spatial settings by summoning nostalgic connections that are either embodied or refer to corporeal connections to the outdoors.

During fieldwork, I spoke to a local seeking some sun during his lunch break. He maintained that many citizens find nature meanings in the OOH's architecture precisely because of their tight-knit relations to it. He painted his idea of the typical Norwegian landscape made up of: 'small cities, and mountains are all around, and sports and leisure everywhere'. This points to an interesting dichotomy between the Norwegian outdoors used as a space to explore, and one utilised for competitive sport. These two sides of Norway's nature relations come with very different values and meanings and are also organised differently by different bodies at the institutional level (Gurholdt & Haukeland 2020, 167). In my time observing the OOH both were present, which in terms of movement relates to both an opening up to, and a bounding from, the world.

During the peak of the COVID pandemic in the summer of 2021, the OOH's foyer was running limited opening hours. No productions were on show and travel restrictions meant that a majority of tourists were Norwegians forced to spend their vacation on home soil. However, despite a poor spell of weather, the roof remained a major attraction. The most noticeable difference during this particular week was the abundance of electric scooters and how they were driven. From my fieldnotes, I recall:

> The foyer is closed. A gale is blowing but it's fairly warm. Several groups of people return from their roof walk all smiles and laughter, exhilarated by the windy conditions up top. Scooters zip down at high-speed utilising the open space. Several riders don't just go over the roof, but make a circuit, up and down, returning to make the descent again, filming each other on mobile phones.

A later entry recalls: 'so many scooters. A young man rides down the entire steps of the slope, filmed by his companions. Another scooter rider almost falls off when he hits one of the gaps in the stonework'. The scooters were a contentious issue at the time, but during this extraordinary pandemic period, fewer visitors meant those present had more space to explore the city. Norwegian groups and families could be seen interpreting the mountain literally, taking to the OOH's slopes as if they were dry ski runs. At this time, there was a distinct 'looseness' of space where people were navigating the rules of what was considered acceptable behaviour, producing and experiencing space simultaneously (Franck & Stevens 2007, 14). The loosening of space is a feature of the roof,

Figure 5.5 Scooter riders appropriate the roof as a utility space. Photograph by the author.

but this particular loose atmosphere was something of an anomaly, in that it was the result of specific circumstances. My own perception was that it held a tension in that people were traversing the space and defining for themselves what was apt. Re-opening from COVID and an influx of Norwegian tourists, combined to (re)navigate the rules of the roof. In a sense, visitors were picking up on the OOH's promise of openness, exploring the access and tolerance of the mountain in the city, appropriating the space through the practice of scooter riding.

Upon returning to Oslo months later, driving scooters on the OOH has since been more firmly discouraged. An employee told me: 'when the electric scooters were new it was like a playground for people, so we had to stop them'. However, an activity like scooter riding is clearly embedded in the way Norwegians participate with the outdoors. The interpretation of the roof as a sports utility was not what the architects had in mind; neither is it encouraged by the Opera company but in shaping an open and loose atmosphere in connection with the outdoors, some Norwegians take the mountain literally. This interpretation contrasts another Norwegian concept, one close to the idea of a 'social monument', that of outdoor living or *friluftsliv*.

Friluftsliv can be defined as 'a household word for a whole range of non-competitive and non-motorised activities in nature, such as hiking, paddling, camping, and angling, with recreation, nature experiences, personal pleasure, and adventures as the main purposes' (Gurholdt & Haukeland 2020, 168). To draw parallels with the different forms of movement discussed thus far, a friluftsliv perspective speaks to the more motile act of roaming, spending time in and with the land. By chance, I interviewed a local woman who was an educated practitioner of friluftsliv. She explained the concept of historical importance as a 'national anchor to bring everyone together', particularly at a time in Norwegian history when the country lacked social cohesion. Most poignant to our interest in movement as a mode of participation is the connection between friluftsliv and trekking that binds movement to ideals that stretch beyond the act of simply walking the OOH's roof. It is these understandings that informants express through their perceptions of Norwegian landscape on the roof.

Trekking as a highly accessible form of interacting with the Norwegian landscape is largely the result of its promotion as the primary outdoor activity by the Norwegian Tourist Association (DNT). The DNT, established in 1868, intended to develop tourism in Norway whilst providing access for all Norwegians to its natural areas (Ween & Abram 2012, 157). As a result, trekking today holds certain political imperatives, leftover from a time of dispute around citizen's land rights, eventually culminating in the passing of the *fjelloven* or mountain law, thus securing the Norwegian countryside as a shared common. This, in turn, led to the friluftsloven being passed in 1957 (Ween & Abram 2012, 160).

The claim that Norwegians navigate the roof in identifiable ways, 'walk[ing] cautiously and with control' (Hammer 2011, 93), whilst speculative makes us consider the cultural codes embedded in people's movement. For Norwegians, these codes relate to an understanding of friluftsliv as an 'outdoor-life ideology, policy focus, and everyday way of life that engages a diversity of skills and types of learning' (Giulianotti 2020, 238). It evidences that Norwegians walking the roof attune to a collectively definable atmosphere. And whilst the roof appeals to a universal experience of the outdoors, for Norwegians there are additional layers of cultural memory and nostalgia, inaccessible to tourists and their general understandings of the Nordic.

The act of trekking has been culturally shaped over many years, through the comprehensive construction of an infrastructure for outdoor experience, 'cabins, shelters, and marked trails', installed by the DNT to provide access to nature and an ethos of democratic intent (Gurholdt & Haukeland 2020, 172). This intervention in the Norwegian landscape can be read as a type of smoothing that has removed a majority of the unexpected and obtrusive elements (Wilk 2005, 23). Likewise, the OOH presents a 'veneering' of space (ibid., 27), a smoothing of the journey that encourages some elements of Norwegianness over others.

The DNT specifically promoted the idea that it is through the use of nature that people become equal; 'we are all the same in Nature' (Ween & Abram 2012, 165). This egalitarianism through interaction with nature was explained to me by Ella, a local woman I spoke to on the roof: 'nature is very fused into the Norwegian identity, and the feeling of being Norwegian' she explained to me. Such sentiments speak to the central concepts that interviewees often express, revealing tones of proud Norwegian nationalism that act for the common good.

To qualify this further, consider how the infrastructure made by the DNT requires individuals to use and maintain it. Trails are not laid on as a service but require the active participation of people. The term *dugnad* infers an egalitarian vision of collective participation in shared tasks. People are voluntarily active through trekking and are encouraged to take ownership, but under the responsibility of 'performing the nature that the DNT has brought to Norwegian nationalism' (Ween & Abram 2012, 166). Freedom under responsibility and Norwegian ideals such as dugnad become part of the OOH's atmosphere that emerges through movement, understood as trekking in the outdoors. And participation through movement therefore becomes constituted not by signs and guides, but by 'seemingly invisible cultural codes' (Urry & Larsen 2011, 192). Furthermore, these cultural codes enhance the relationship locals have with the building, as one tour guide suggested when saying: 'I think I speak for other Norwegians, you have a relationship with the house, you feel proud, and have an ownership to the house which doesn't have anything to do with the artforms'.

Perhaps it is no coincidence that in present day Bjørvika a walk from the city over the OOH leads to a new DNT friluftshus. Here locals and visitors can sign up for onsite activities, presentations, or be guided to walking tours outside of Oslo (Friluftshuset 2021). With its flow of new arrivals and access to water activities, the DNT continues to promote its particular blend of Norwegianness and tourism, reproducing what has been described as a 'Norwegian sense of nature-nationalism' (Ween & Abram 2012, 157).

Resonance, dissonance, and a good-natured elitism

Every year, millions of people flow through the OOH and its roof area. When we zoom out the building appears as a mobility hub, a nodal point, pushing and pulling visitors from the city to Bjørvika with such success that a detour over the roof has become an unmissable part of the Oslo experience, practised by tourists and locals alike. That the OOH can hold such ambiguity, shaping universal as well as local relationships between people and architecture is, I suggest, partly the result of the peculiar forms of movement it engenders. Moving on, the roof connects people to place in very personal ways; they attune to the environment through their body. This is individual in essence but, as I have argued, can also assemble collective understandings and shared ideals. Up close, the design of the roof forges subtle conduits, staged atmospheric transitions that engage visitors, pushing them around the OOH in pursuit of new discoveries and remarkable vistas. Turning to unforeseeable events and 'mishaps' as a part of an attunement to the OOH, movement on the roof shapes an atmosphere of potential and futurity that is both designed and transcendent of what design is able to determine.

Exposure to weather and the elements heightens people's experience on the roof, shaping embodied encounters and underlining how motile movement can shape authentic and meaningful experience. Paradoxically, mishaps can lead to feelings of inhabitation as attunement through design becomes re-worked by the way people participate through dissonance, forming unique relations between architecture, individual, and world. Mishaps such as tripping and encounters with non-humans such as gulls show dissonance to be essential to making connections with the world, as it is here that the excesses of atmosphere can be found spilling over.

From a design perspective, some kind of balance must be achieved. That is to say, there is a point where mishaps become more than atmospherically dissonant and experience becomes wholly negative, for example, somebody becoming physically injured. Extreme dissonance, as we discussed in a previous chapter with regard to glare and excess light, erases atmosphere. What is essential is that individuals feel an underlying potency for new experience, an unrealised potential that drives them forward. Identifying the feeling of pregnant possibility highlights a way for architects to hold their designs open and charge spaces with an atmosphere of futurity.

For non-locals, the universal experience of the outdoors remains. Most visitors have been to, or are aware of, comparable waterfront buildings. Icons from Sydney, Bilbao, and Copenhagen are notable by a perceived association with the architecture of the OOH. Tourists also carry knowledge of the Nordic countries, nature motifs, and visions of landscape that align past, present, and future. The rendering of nature archetypes therefore arrives through the building's immersive qualities, what Böhme has called the social and synaesthetic characteristics of atmosphere (2017, 60). In turn, atmospheric characteristics seem to lend themselves in different ways to local and global understandings, for example, the synaesthetic relates to individual experience *with* weather, whilst social understandings speak to a notion of climate and a general understanding of the weather conditions associated with a place, or country, that do not necessarily arrive through sensuous experience and memory.

Considering how participation through movement is part of an ongoing formation of Norwegian identity reveals 'a tension between repetition and originality' (see Bergman 2010, 79) as locals participate in doing and re-doing Norwegianness. The invitation to participate implies the potential for individuals to find their own path, but movement, as we have seen, involves more than physical freedom and constitutes specific cultural understandings. Thus, movement shapes an ambiguous opening and closing of meaningful experience, the driver of which is a Norwegian variety of elitism, grounded in specific ways of relating to nature and the outdoors (Hammer 2011, 93). In this manner, the OOH makes disciples of people as they move forward. They advance, believing it to be the best path to take.

Notes

1 Despite its location, the Sydney opera was designed by Danish architect Jørn Utzon. Various disputes led to Utzon leaving the Sydney Opera House project prematurely. In fact, he never saw his nature-inspired masterwork in completion (sydneyopera-house 2023) (https://www.sydneyoperahouse.com/our-story/utzon-departs-the-house#:~:text=But%20Utzon%20was%20never%20invited,more%20perfect%20for%20remaining%20unrealised.).
2 Graphic design uses the term ‘proportional density’ to describe the way a design conveys memorable and meaningful information (Lidwell et al. 2010, 190).
3 The station commons (stationsalmening) is the area adjacent to the Opera house that also holds Oslo’s main train station.
4 A ha-ha is a ditch that is invisible from distance, allowing different spatial zones to be divided but without disrupting views (Janson & Tigges 2014, 136).
5 I heard this adage several times from different interviewees. For an amusing outsider perspective, see: https://www.lifeinnorway.net/norwegians-born-with-skis-on-our-feet/. Accessed 10/2021.

6 Light participation

I follow the white marble 'carpet' that continues from the Opera's outer areas, entering the foyer space through one of three rotating doors. The sound of the city fades away, replaced by the whooshing sound of stiff brushes against a hard floor. I pass under a low overhang and enter, pausing when confronted by the sudden change in scale, surprised by the height of the ceiling and a feeling of open space. The acoustics have changed but the strong natural light remains, enhanced by the marble that has received a smoothing treatment giving the classic stone its familiar luxuriant quality. Two young girls run and slide up and down the hall in their socks, demonstrating the stone's lustrous material characteristics. The room is predominantly white except for two eye-catching facades on each side of the space: a hue of candlelight orange-tinted wood on the left and the cool green-white glow of a ceramic-looking wall to the right. Looking up, the huge glass apertures present vast screens of sky in grid form. Angled pillars are sporadically placed, crossing the up-and-down geometries of the room with diagonals. A man appears with a camera device on a long pole, capturing the space from extraordinary perspectives. I turn towards where I have come from, noticing the glow of spotlights embedded in the structure of the entrance. Their hidden fixtures give a metallic orange-pink hue that exudes from the surface of the material. The entire space is alive with light that appears to have no source. A youngster with a scooter enters, stopping by the entrance. He rides in a half circle and leaves again. People walking the roof are visible in glimpses, sometimes stopping to peer into the space from above. A couple enters with a dog. One of them waits with their pet whilst the other moves towards the toilet area. A group of youngsters sit by a bench but use the floor forming a messy circle, nobody seems to mind. At the south end is a café restaurant and further down, behind the curve of the wood wall, lines of tables draped in white cloth seat a few diners for lunch. Beyond is water, the fjord, and the occasional seagull that zips by on the wind. Waiters and staff walk through the space at tangents carrying out their tasks and a lone security guard ambles slowly around. Taking the middle of three short passages that lead behind *The other wall*, we enter the cloakroom and toilet area. The acoustics change markedly. Pressed into the sharp angle of the building's west side, two kids and a woman use the marble incline to slide down, shrieking with

DOI: 10.4324/9781032668802-6

Figure 6.1 A composite of photos taken inside the OOH's foyer. Photograph by the author.

laughter that bounces around the space. Another family sits on a bench talking, their voices also carrying and reverberating. Moving away from the noise and intimacy of others, I take a seat at a table in the foyer. Two men enter, one quickly removes his shirt and inverts himself to hold a handstand whilst his associate lies on the floor and, using the wood wall as a backdrop, takes his picture. The activity gets attention from the guard who moves in to check their credentials, seemingly encouraging them to desist. The photographer seems adamant, waving some papers without rising from his position on the ground. The guard appears unimpressed but retreats enough for the pair to quickly shoot a few more images before leaving the space as quickly as they came.

Nordic light, Nordic architecture: from the roof to the foyer

In the previous chapter, I have looked at movement and the way people inhabit the OOH through its material design, surrounding elements, and a Norwegian sensibility to the outdoors. I have also previously interrogated marble as an atmospheric material and constituent of perceptions relating to nature, landscape, and culture. Participation relating to movement and materiality inevitably involves light due to its role as a shaper of social and material life (Bille 2015). Acknowledging its relational character, here I focus on light as an atmospheric element and mode of participation that is complicit in the staging of experience at the OOH. Taking into account the differing forms of light: natural, artificial, and the transitions in between, I argue how light at the OOH produces an atmospheric continuum, a staging of openness and possibility that locates the OOH as an inseparable part of Oslo's contemporary urban development and greater geographic zone.

Thinking atmospherically, the presence of multiple light sources implies a diversity of experience and, therefore, holds the potential for numerous atmospheres to be created as individuals form ever-new cominglings with their material and immaterial surroundings. However, people's general perception of the OOH is found to be one of cohesion and homogeneity. This chapter approaches this paradox by engaging with two main theoretical perspectives. Firstly, Gernot Böhme (2017) and his writing on light as a key element and generator of atmosphere. And secondly, Peter Sloterdijk (2014, 2016) whose concepts of spheres and foam I bring into dialogue with Böhme's notion of atmosphere as the co-presence of subject and object. By my reading, Sloterdijk's theories allow us to nuance atmosphere as a subject-dependent atmospheric space, transcending the specific surfaces of architecture and drawing attention to its connection with its greater surrounds via a conditioning of the air (2016, 166). Following the idea that light and its transitions emphasise the temporal dimensions of atmosphere (Dalsgaard & Kortbek 2009, 4), my intention is to explore how light makes the dividing membranes of space more or less porous, between what Sloterdijk would regard as different atmospheric spheres. Of particular interest is the way light mediates the borders of the OOH's built, symbolic, material, and immaterial space, manifesting notions of *Nordic light*.

Following a consistent theme of this book, I inquire into how light as an element of atmosphere is complicit in the smoothing of ambiguous terms. How a variety of simultaneous themes and functions co-exist, and how the tension between democratic space and the building's role in placemaking, tourism, and landmark iconography is mitigated. Thus, I build outwards from situated atmosphere to the social foam of experience, explaining how the homogeneity of the OOH's atmosphere is co-constituted by light.

Entering the foyer, light becomes visibly present through its interaction with the other elements in the space. Surfaces and windows dynamically shape the relationship between natural and artificial light, the inside and outside, building and user. Notable too is the diversity of light from an assortment of sources that surround the foyer's activities. Most prominently, large banks of windows allow the outside to penetrate the space as the sun swings across the sky from the building's easterly side, resulting in shifting moods dictated by Oslo's weather conditions. Natural light becomes entangled with the artificial lighting fixtures, for example, the *wave wall* and *The other wall* which dominate the space as daylight fades, changing the OOH into what Snøhetta call a 'lantern'. In this manner, light creates a tidal movement of changing atmospheric effects, a diurnal rhythm where the OOH sucks in the natural light from outside, then turns to exhale it back to the city. Metaphorically, we might describe this process as moving from *camera obscura*[1] to lantern, shaping ever-new combinations between the architecture, city, and environment.

Figure 6.2 *The other wall* and the oak-clad *wave wall*. Photograph by the author.

Böhme draws our attention to architecture as a sounding board for the connections between local climate and modes of participation, showing how atmospheres are co-constituted by weather (2017, 79). Light as a natural phenomenon makes the membranes between architectural boundaries porous, dematerialising the material and shaping environmental as well as existential encounters (see Bille & Sørensen 2016, 2). Snøhetta talk of such membranes as points of transition which can be used to direct people through interesting assemblages of light and tectonics. In public spaces, such arrangements are often navigations between design and everyday practices, for example, upon returning to the OOH after some months, I found the foyer shop had moved location. During a conversation with an employee, I was informed:

> the previous shop area was not actually made for the shop, because you must have free sight to the Eliasson, that's why it moved. Snøhetta hadn't told us their plans at first, then they told us and that's we are moving the shop.

Free from the shop's intrusion, light from Eliasson's *The other wall* was visible from the outside, its subtle changes observable from the street.

From the transition from camera obscura to lantern, it is implied that unique moments between individuals, light, and architecture emerge. For Böhme, the presence of different people with differing cultural interpretations suggests the indeterminacy of atmosphere as there is always potential for different atmospheres to be shaped (1993, 114). The possibility for new and diverse atmospheres suggests their presence to be a gauge for diverse forms of participation, for without difference, hegemonic perspectives are enforced rather than challenged (Cornwall 2002, 5). In other words, the foyer's feeling of openness and potential perceived by most visitors lends itself to a spatialised form of democracy and participation. Light as an atmospheric element becomes a vital part of the OOH's participatory propositions, promising an encounter awaiting activation by the subject. From this perspective, the play of light shaped at the OOH becomes a means of engaging the visitor and a gesture towards participation. Individuals are drawn into 'the self-evaluative quality of experience', *seeing themselves seeing* (Eliasson & Irwin 2008, 55). They become conscious of themselves perceiving a place in a certain way that reveals the conditions of their experience, and this is taken to be an act of self-empowerment. There is a link between such participatory claims and the Nordic approach to development that places such an onus on the social. Nordic governments profile themselves as global frontrunners in openness and transparency (Marklund 2014, 249) whilst Nordic 'builders are expected to plan for social spaces', to 'think about everyone's access to views, daylight, and a social life' (Tingsby 2016, 56). As architecture inevitably includes political ideals, light becomes a tool for their manifestation.

The Nordic climate with its long summer and short winter days has traditionally defined people's social activities. As outlined earlier, Snøhetta were open about taking advantage of the Norwegian response to nature and landscape, actively incorporating 'Nordic think' into their designs. In this sense, light at the OOH relates less to a 'universal fact' (Le Corbusier in Hawkes 2020, 67) of the sun's path moving across the sky, and everything to do with how it is made to feel atmospherically.

We know by now that Nordic architecture often aspires to genius loci. However, the *spirit of place* brought into resonance with the OOH is not just its immediate surrounds but pertains to imaginaries of the wild far north. As the outside is allowed to enter the OOH's interior, visitors attune to a Nordic outdoor which, as argued previously, is understood to relate to specific cultural meanings. Light becomes Nordic light as Snøhetta, 'pull on the heart strings' of locals, attuning them through their embedded sensibilities: 'one can literally feel the presence of Nordic light, be it daylight or artificial lighting' (Haug 2011, 1). It exemplifies how light, and lighting, is not simply reliant on our perception, but is culturally connected to specific places and activities (Sumartojo, Edensor, Pink, 2019, 3), capable of producing decipherable meanings.

As I shall unfold, a key aspect of the OOH foyer relates to its porous borders that allow a Nordic outside to enter its physical interior. The permeability of inside and outside conveys notions of transparency and equality, grounded in a notion of democratic and participatory space that invites us in (Cornwall 2004, 2). At the same time, Nordic light is incorporated into the OOH's architecture becoming a branding tool universally understood by visitors, as well as a subjectively perceived phenomenon shot through with cultural understandings.

It is in these tensions between openings and closures, local and universal, inside, and outside, that I explore light as a mode of participation.

Daylight in the OOH

I enter the foyer on a rainy day in July. It's unusually cold and changeable. Inside the foyer, I buy a cup of tea and ask about closing hours. A waiter informs me that under the current pandemic conditions, there is flexibility around closing time and it generally depends upon the weather: 'it's raining, so everybody is in the centre of town'. By everybody, he was referring to the summer tourists who would otherwise be flocking to and through the OOH: 'Last week', I was assured, 'it was hot and very busy'.

Later that week, I overheard a guide talking to his group asking how they felt about the space. Quite unanimously they reply: 'it feels open and light'. Somebody commented that they were also here last winter, and it even felt light then. The guide mentioned that the architects deliberately used as few materials as possible, using the term, 'expressionistic minimalism'. A member of the group introduces himself as a carpenter, saying that upon entry he just stood and marvelled at the pillars: 'the columns are so clever, to hold all that weight and yet provide an open space'. Such designs increase the flow of light by removing large blocking elements, meaning that in certain conditions the space becomes dominated by the natural light from outside.

Moving through the low overhang and entering the foyer from the north side, the abundance of light and space momentarily halts visitors. We have reached another of Snøhetta's transition points. This type of visual stimulation encourages people to wander the room with their gaze whilst the distance between the body and all other objects gives a sense of security as a space without objects can be explored with freedom (Böhme 2017, 149). This can be exemplified by the many children who slide or run into the space, enjoying the chance to move in uninhibited ways. Such moments hint at the way the quality of light shapes the felt perceptions of the space. Lightness refers to an abundance of light, but also to an atmosphere of openness and weightlessness, as Ingold notes, atmospheres can straddle the meteorological and aesthetic, the cosmic and the affective (2016, 168).

Weather is an intrinsic part of our everyday lives as the atmospheric state of a particular place at a particular time (Retallack 2004, 198), a point also noted by the American architect Lois Kahn who once remarked that allowing natural light to enter a building produce, 'as many moods as there are moments in time, no single day like the other' (1971, 16). It is these moments that designers can seek to enhance, allowing visitors to become part of the 'magic' that comes from 'projecting oneself, as it were, into these light spaces' (Böhme 2017, 150).

In Nordic countries, the connection between place, landscape, and national identity gives light a privileged role (Edensor 2017, 36). People's relationship to the outdoors is particularly poignant due to its extreme intensities, meaning 'weather and feelings are closely related' (Böhme 2017, 69). Inside the foyer, the *use* of natural light follows the natural variations of seasonal change, creating a flow of moods. Unlike humidity or temperature, light literally enters architectural space, albeit through subtle filtrations and staged openings. The OOH's large apertures and use of light-sensitive materials (Vollsnes 2010, 28) tell us light was a major consideration during the building's conception. Placing the OOH over the water brought the realisation that a freestanding building would inevitably interact with the fjord, and natural light would heighten the building's connection with its environment.

In all seasons the sun lies approximately to the south of the OOH, but its changing elevation over the horizon, its solar zenith, creates differing angles of light that interact in distinct ways with the building's materiality, altering the passage of light through the foyer. In December the sun rises around 9:30 am, meaning those present will experience low sunlight from the east that will move south around the fjord, and then over the water around noon. A characteristic of light at this time of year is its low elevation, never rising more than eight degrees above the horizon at its peak. This winter light can be virtually horizontal, even appearing to arrive from below the horizon (Plummer 2012, 6). On a clear winter day, the low sun enters from the southern side through the café area, throwing long shadows into the space. The pillars create heavy stripes that can fill the entire length of the foyer with their shadows, slicing the space into sections. People interact with this phenomenon noticing the dramatic combinations of architecture and natural light, what Franck and Stevens refer to as a playful, soft, or loose threshold (2007, 73). From my fieldnotes, I recall how: two women sit on the floor, one photographs the other experimenting with various angles, directing her subject into different parts of the space defined by the strong pillars of light and shadow. Light in this state recomposes the building's form, prompting people to move and use the space differently.

Böhme states, 'darkness is, in the interaction with light, the condition for seeing something' (2017, 148). In other words, darkness articulates visual boundaries. Here, darkness in the form of shadows disrupts the OOH's design, de-materialising its form. Inside the foyer, when experiencing an oblique sun, guests move to follow the temporary new paths that light has carved out for them. Relatedly, a visitor expressed how the immersive qualities of the light have a bearing on social practices, relaying that: 'in a way, light shapes what you do and how you spend your time here'. Notably, visitors position themselves according to the light and not the architectural space, often drawn into exploring areas that would otherwise be overlooked. Much of their participation involves photography, playfully capturing the dynamic elements of the space and its collaboration with light and shadow. Reviewing Instagram posts of the OOH, these moments of phenomenological encounter are exaggerated through framed fusions of light and materiality, high contrast, and saturated colours that further enhance what the visitor has experienced.

In July, the sun rises from the north-east at around four in the morning, setting around 22:30. The high sun begins to enter the foyer around noon, then moves south throwing shadows onto the wood wall. People and gulls moving on the roof can be seen in silhouettes, projected onto the walls. At this time of year, the angle of elevation over the horizon is around 22 degrees. Sitting at a table in the foyer on a summer's day, I observed the quivering mottled refraction of light from the water of the fjord bouncing onto the ceiling. People entering the space would stop up to observe the dancing light overhead, wondering at its source.

The communion between building and light produces incidental encounters that vary with the seasons, shaping intricate light plays. The changes in light conditions alter the material feel of the space as textures and contours become altered. In such moments, the OOH can be embraced as a kind of *Aeolian harp*[2], an instrument that gathers elements from the outside, translating them into 'the music of the cosmos' (Pallasmaa 2011, 27) and a resonance that suspends immediate cultural meaning in lieu of a felt connection with nature and a romantic notion of the outdoors.

Just as seasons change, so too does the city and Bjørvika. During my visits to the OOH, I observed the light effects caused by adjacent buildings such as the Deichman Library. New constructions bring new tricks of light, as local buildings come into spatial dialogue with the OOH. Over time, the sun touches their respective surfaces and makes them part of an experiential process where architecture summons our imagination, attention, and movement (Plummer 1987, 75). As I will later discuss, transitions between natural light and the foyer's artificial lighting interact dynamically, conjuring a technological re-creation of the experience of nature as an ecological spectacle (Sloterdijk in Löschke 2016, 4). By directing flows of natural light, the foyer shapes Nordic light's peculiar qualities, distributing sunlight and drawing our attention to the

Figure 6.3 Silhouettes of activity on the roof, cast upon the wooden *wave wall*. Photograph by the author.

strength and angle of the sun, in combination with cloud and dust, shaping ever-new diffusions of light (Edensor & Hughes 2019).

Whereas buildings were once typically solid and opaque, glass provides visual transparency that speaks to a tacit ideology. As Foster writes, 'many architects have associated this disclosure not only with professional honesty but with democratic openness' (2013 98). Through the high windows, sections of sky are observed from the foyer. People generally acknowledge the predominance of daylight and perceive the space as unusually in tune with the outdoors. The foyer's glass and its pellucid qualities give multiple perspectives, merging the light from distant forms, dispersions, and reflections (Plummer 1987, 27), that entice by way of a fracturing of inside and outside. In this sense, glass is notable as a material with strong immaterial qualities that follows a Nordic architectural tradition where light is not used to decorate, but to transform and invigorate (Plummer 2012, 100). Apertures become screens, architectural events that display points of change between the outside and the inside which people know to be different (Venturi quoted in McMurtrie 2017, 39). There is something in this knowledge of artifice at play that bestows whatever is being framed with extra vitality and interest. Böhme writes about windows as 'light like objects', that have a spatial presence not as a light source but as ecstatic things in themselves (2017, 150). Glowing, these objects proceed into the space, transcending the rational and encouraging the viewer to enter the realm of the sensual (Edensor 2015a, 139). Despite, its green tinge, the bank of glass in the foyer gives centre stage to the natural light that pours in. The subtly tinctured glass provides just enough protection from the blinding glare of the sun, yet remains concealed from those inside becoming, 'sky, phenomenology, filtered reality' (Philippopoulos-Mihalopoulos 2016, 154).

On fine summer days, the doors to the south are thrown open, the breeze from the water can be felt, and the light and sound of the outside enter the foyer. It is at such times, when the membranes between inside and outside are punctured, that the staging of the senses becomes most apparent. But even when the borders to the outside are sealed, the notable lack of artificial lighting shows a reverence to the outdoors and a close association that displays the passing seasons.

Allowing light to enter the foyer space brings with it the varying moods of weather and a general atmosphere of openness, conveyed through the staged fragmentation of inside and outside, and the mediating qualities of light which rupture the foyer's material borders. Light contributes to the presentation of the foyer as a unique and 'quizzical' space, precisely because it allows a dialogue with nature-like experiences. Moreover, we see that the outside that penetrates the membrane of the foyer is not only Oslo's local environment, the city, gulls, and traffic but a Nordic outside, derived from the far north, that resides in the imagination of visitors. Thus, in the space created by the Aeolian harp, a culturally embedded vision of the outdoors and a specific sensibility to nature is asserted.

Artificial light

With its extended durations, contrasts, and soft diffuse qualities Nordic light is considered as 'something special' (Madsen & Kristensen 2006, 94). Its place in the cultural heritage of Nordic lands has changed the way Nordic architecture has used light as a medium, 'tempered by the moody Nordic sky' (Plummer 2012, 184

As stated, during the day people in the foyer tend not to see light, only sky and weather as the outdoors is gathered within the OOH. One bright summer day I spoke to an employee on her lunch break. She noted: 'the light does not catch your eye so much in the day when its more natural' but assured me 'it becomes more dramatic at night'. Oslo's northerly 59.9 degrees longitude gives a huge discrepancy in the arrival of darkness at the OOH. In summer, anywhere north of 61 degrees night never falls (Garnert 2011, 6), whilst at the OOH an extended dusk and dawn buffers a short-lived sundown. In contrast, winter brings extensive periods of darkness, days where the sun barely seems to rise. It is in these wintery states where it becomes particularly pertinent to apply Snøhetta's description of the OOH as a lantern.

From outside the OOH, what we see emanates from a warm light touching the wood wall that envelopes the Opera stage's exterior. Light seen from the outside can carry cultural and social meaning, as Bille writes, light sources such as candles and lanterns in domestic settings give a sense of 'being both inside and outside at the same time [..] extending not only the apartment but

Figure 6.4 At night the OOH becomes a lantern. Photograph by the author.

also oneself as a person, as a part of the city' (2015, 62). This, 'shaping [of] social closeness through light' (ibid.) can be applied in relation to the OOH as it turns itself into a welcoming beacon for the city. Relatedly, Norberg-Schultz describes Norwegian *place* as founded on a tension between above and below, a reflection of living in valleys between mountain peaks. The orange firelight of the shelter is in tandem with the harshness of thriving outdoors. Norberg-Schultz argues that 'to reveal and maintain this in building requires forms that simultaneously possess the safety of home and express the indefinite and savage environment' (1996, 36). Just as Bille describes a homely atmosphere that transcends its borders with a social light, the OOH as a lantern emits a social light to the city.

Having entered the lantern, people tend to move through the centre of the space between the two installations. The route immediately to the right leads to the toilets and cloakroom, so functionality plays a role as people disappear behind *The other wall*. The doors act as another transitional element, their swoosh crossfades the sound of the outside with the interior, coinciding with the visual contrast of the foyer's scale and tectonic features. People often stop, just as they do when crossing the bridge from the city, taking in the new space and deciding what to do next.

Aside from the two installations, the drama of the foyer's mood comes from its subtle use of *dark light*. As a lighting designer from Oslo explained: 'good lighting consists of layers, there is usually a ground layer of functional lighting, a brightness that allows us to move around safely, and then a joy factor, a brilliance and colour that makes you think about things'. In other words, lighting design conceives of a space as a collection of separate parts and elements. It has a functional basis, allowing people to see, as well as an experiential, emotional response. Clearly, this is where atmosphere becomes pertinent to architecture, relating to a trend in design that recognises light's influence in the shaping of our urban environments (Mansfield 2018, 87).

As the daylight fades, the up-lights and down-lights built into the foyer's interior make parts of the building visible. Snøhetta avoided lighting fixtures:

> most of it is reflected light off a surface, we have tried to hide as many lamps as possible and that is part of the process of changing the colour tones and the scale of the building, changing the tactility of the spaces.

Unlike the use of lamps which we understand to be producers of light, reflected light opens the space and softly illuminates the building as a background for the more prominent lighting installations to overlay a 'joy factor'.

Dark light's interaction with the foyer's architecture causes its materiality to take on a more immaterial manifestation. Instead of hard-focused light that picks out the contours of material form, dark light doesn't hold objects, 'instead it fractures the built form into tonalities and shades' (Plummer 2012, 18). Distinctive from the 'harsh, immediate and technical light of the light bulb', we find a 'mellow' lighting that conjures comfortable 'domesticated'

Figure 6.5 The cold tone of *The other wall* is contrasted by the warmth of the wood wall. Photograph by the author.

inhabitation (Pallasmaa 2011, 29). Such cosiness is not recognised by all visitors, as one expressed: 'this place is not *hygge*, its more about experience. You don't come here to cover up in blankets and drink hot chocolate. No, it's more a space that gives you something different than your home'. Despite its Nordic associations, the OOH as an extraordinary space is always held just beneath the surface of people's perceptions. And, as with the natural light outside on the roof, the interaction between materials and light shapes novel encounters, bestowing parts of the building with a temporary aura.

At night, and with dark light in the background, the foyer is dominated by its two main installations: the *wave wall*, designed by Snøhetta, and *The other wall*, Olafur Eliasson's art come architecture piece. The installations face off, contrasting each other through their use of materials and colouration. The wood wall provides a stable and attractive warm glow whilst *The other wall* is a cold white and green porcelain that pulses in slow rhythmic cycles.

A group of four young students summed up this opposition as they sat on the floor in front of the wave wall. They appreciated the way they could just come and spend time in the space: 'it's cosy here' they explained, pointing to the fur benches that are placed in front of the wood wall. Henry Plummer writes that an intimate relationship with wood is evidenced by traditional Nordic huts and cabins. Derived from its rural history, wood is 'considered a living material, permeated with myth and folklore' (2012, 120). The choice of materials, wood, and animal skin, offsets the marble, glass, and the cool ceramic tones of *The other wall* to provoke a feeling of cosiness and sense of community (see Bille 2015) that speaks to a Nordic aesthetic. A foreign student I spoke to in the foyer recognised this 'Nordic feel, the design, the colours, like a Nordic landscape, white snow and brown trees'. In this respect, it is easy to see the inspiration for Snøhetta's wave wall, just as it is recognised by visitors as an essentially Nordic feeling installation, and it is light that brings forth the qualities of the wood essential to eliciting homely candlelight tones.

From these observations, the wave wall can be regarded as a form of what Sloterdijk calls an architecture of immunity (2004), enforcing an understanding of Nordic materials and light against an outside. It follows then that the students sitting on the floor perceived *The other wall* as follows: 'it's cool but doesn't fit in with the rest of the place. Yeah, the green light, it's weird but I do like it, it's spacey. Like outer space'. Several others also agreed that *The other wall's* aesthetics were 'unusual', questioning its choice of green light.

Interestingly, the contrast between the lighting installations is owned by the architects as a deliberate strategy for manoeuvring people around the space, explaining:

> I'm feeling I am in a cold area, its white, its sharp, the light is sharp, it has this Nordic light quality to it, this cold feeling to it, and there is this fantastic wooden wall which is warm. I mean tactilely warm, but also, I carry that warmth in my brain, and I know that wood is warm, I know it's a different kind of thing to touch, its softer in some way, so I'm gonna go towards the warm bit, it's human instinct.

This tells us that two contrasting atmospheres have been staged, one that pushes people away with its felt coldness, and another that draws people to it. Amidst this tension, how does an overall perception of openness and harmony become maintained? The architect's account requires nuancing. Firstly, benches line the wave wall, providing an additional motivation for people to move towards it. Secondly, the architect is referring here to the process of leading people to their seats in the opera, where they must pass by and through the wave wall. Clearly, the functionality of the opera distorts people's use, especially for those with a ticket. From my own observations of people in the space, the 'weirdness' of *The other wall* attracts visitors to touch and photograph it with greater frequency than that of the wooden wave wall. Its otherness seems to compel people to explore it in a different way than the familiar Nordic tones of the warm wood. A guide supported this, confirming: 'visitors do a lot of selfies here, the Eliasson is very popular, you get very nice light, I have one myself'. If the wave wall presents a familiar and cosy Nordic encounter, *The other wall* offers a novel curiosity where light is subtly changing the material appearance of its facade.

The number of individuals photographing *The other wall* is higher than the other parts of the foyer and it was observed that people would often take several pictures from different angles and locations. Many would not move far beyond the entrance area, stopping near the door and taking a general picture of the entire space before moving to Eliasson and ending with a *selfie*, before leaving. And yet, when reflecting upon their experiences people tend not to differentiate between the different atmospheres that the two lighting installations create. Applying Sloterdijk's thinking, we can think of the foyer not as contrasting spheres but as spheres in bounded relation with one another, existing in 'co-isolation' (Sloterdijk's 2016, 56).

During my conversation with the four students sitting on the floor of the foyer, they observed how balanced and well-lit the space was. The light is: 'really good, the whole place is just glowing without there being any real lights'. They insightfully observed that had strong lighting been used: 'it wouldn't be nice, it would be distracting, and when it's light outside in the summer you don't need lights'. They realised that the views of the city they currently had from the foyer were due to the design of the light and the placement of materials that allowed an opening to, and resonance with the outdoors.

Interpreting the foyer space as co-isolated bubbles that form 'foam' (Sloterdijk's 2016, 56), reveals how artificial light brings atmospheric spheres from the outside into relation with the foyer. Atmospheric encounters have 'afterlives' (Ash 2015 90) as people move around the architectural space. Atmosphere conceived of as a co-presence between subject and object becomes unhelpful here, precisely because atmospheres are always in dynamic relation with one another. Taking into account light's qualities in the foyer space should, therefore, not be taken in isolation, but as part of the fabric of towns and cities (Mansfield 2018, 80). In turn, this leads me to explore the transitions between light and dark, natural, and artificial light, which shows itself to be essential in understanding light's role as a participatory mode.

Transitions: liquid light

Talking to an employee, they noted the sense of awe that weather can inspire inside the foyer: 'when I am in the shop, I can see it coming, you can see it all around', explaining how the materials resonate with the outside: 'because of the colour, you can see the light in it, you can see the weather in the marble'.

I have already discussed how natural light and the OOH's designed materiality aid the perception of a porous and open foyer space and a general atmosphere of openness. I now briefly explore transitions, understood as the shifting conditions between natural light, artificial light, and darkness. Additionally, transitions are considered as a mechanism that connects the foyer to other local spaces via bodies, moving in and around the OOH's local area. From an atmospheric perspective, I have raised the question of why a perceived atmosphere of openness leads to homogeneity rather than the emergence of multiple atmospheres, which in turn leads me to inquire into light's role in the general atmospheric staging of the OOH. That is to say, how light is complicit in producing a cohesive understanding amongst visitors, when multiple materials, bodies, and spatial zones imply the potential for different atmospheres.

As the sun begins to set in Oslo, the presence of the OOH's artificial lighting comes forth. Aside from the two major installations, the interior lighting is minimal. Only after spending time in the foyer and looking with intent did I begin to realise the source and layout of the artificial light. Instead of using designed lamps, lighting fixtures are well-hidden allowing glows and reflections to come forth and grow in intensity as the sun retreats. The type of 'dark lighting' outlined above can also be seen as a feature of Nordic Light, as a local woman unpacked for me:

> I mean it's so present, it's both the lack of light and the type of light you have, if you go up north the sun doesn't go above the horizon but there is still a lot of light, so it's this other light that isn't warm strong sunlight that's very, very interesting.

The informant, with her local sensibility to light, sketches the ingredients required to identify Nordic light. She continues: 'Nordic light has a lot of transitions in between day and night, very slow, and there is often a lot of orange or purple which you wouldn't have by the equator, so physically, aesthetically, and socially, I think light affects Scandinavians'. This detailed account of light from a reflective citizen was echoed by several others who gave consistent accounts of phenomena such as the *blue hour*, 'a condition enhanced by long dusks and the extreme refraction of the sun at high latitudes' (Plummer 2012, 148), and the general importance of light to Nordic cultural understandings.

Another trend in contemporary lighting design is the use of artificial light to mimic the transitions of nature and its changing colour tones. I spoke to lighting designers ÅF, a company responsible for several other major lighting installations in Oslo. ÅF use the term 'liquid lighting' to refer to programmable installations of light that alter the scale and tactility of material structures. Such techniques extend the designers atmospheric pallet, making light an active agent that when introduced into a space, 'becomes a reality machine' (Eliasson 2014, 93). In this immersive state, light becomes more tangible and emotive, 'perceived as a substance; the liquified light feels like a moist veil on the skin and it even seems to penetrate into the pores of our skin' (Pallasmaa 2011, 29).

Inside the foyer, Eliasson's *The other wall* is an example of sequenced lighting that fluctuates against its surroundings, creating innumerable combinations and interplays between light and materials. Even the constant orange glow of the wood wall, illuminated by single-function fixtures, takes on a liquid quality when placed in a space so exposed to the dynamics of natural light. Indeed, it is often difficult to discern whether the transitions are instigated from within or whether they come from an outside source that has penetrated the foyer's margins. As a result, the liquid feel of the light appears in step with the rhythms of the outdoors, making the foyer's membrane appear permeable.

The use of light has been well documented as a means of enhancing the symbolic qualities of architectural structures associated with state institutions (see Edensor 2017, 90). Numerous architectures exist where light freezes the contours of the building, monumentalising it. It can be argued that when the OOH becomes a 'lantern', it also becomes a form of monument, however, the warm glow that comes from within suggests a welcoming internal activity as opposed to a bounded and static institution. Furthermore, the up-lighting on parts of the building's exterior throws shadows of moving figures on the roof, making it come alive. Thus, the lantern becomes a social light, opposed to the immobilising effects of the monument. Utilising forms of dark and liquid light, the OOH follows the characteristics of Nordic light whilst also enhancing its social credentials.

Through the eyes of the architect the foyer is part of the journey, a slow transition to other experiences. As Snøhetta explain: 'at the end of the journey you are going to be sitting on your own in a small seat in a dark room, looking at something or hearing something'. In this conception, the foyer is a threshold between the roof space and the inner core of the theatre, giving it a prime role as a public space where people and institution most closely relate. But for non-ticket holders, the foyer is just another node on Oslo's greater experiential map, evidenced by the many visitors who enter the space with luggage on their way to or from other destinations. Arriving and departing from the central train station, visitors move through *The sunset*, an award-winning light installation that fills part of the west side of Oslo's train terminal. Its designers ÅF explained how the east-facing terminal used to be a space where light would flood in. After the station was renovated, the light was obstructed, inspiring them to

Figure 6.6 The sunset designed by ÅF. An example of Oslo's many urban light designs. Photograph by the author.

recreate the sunset using 71 independently controlled oversized pixels to create a technological sun (ETC 2021). As with the OOH's foyer, strong light obliterates its effect, paling its colours almost to invisibility. However, in the evening the sunset fills the space with moody tones of colour, presented vertically as an unavoidable luminous object encroaching into the surrounding space.

As a result of Oslo's harbour development, new paths for people to move around the city have opened, and radically changing the way people move around Bjørvika. The frequency of visitors on their way to other areas of the city gives the building something of its unique feel. Having observed several people entering the foyer with bikes and dogs, I asked an employee about what people do inside. I was told that:

> it's quite common for people to enter with dogs as people are on trips walking through Oslo. People also use it as a hangout spot, a public toilet, especially in the summer for swimmers and stuff, and the café as well, so it's quite inviting and it should be like that.

The informant highlights an alliance between mobility policy and the culture-city, one that plots events and encounters along our urban corridors (Sloterdijk 2014, 157). And, of particular relevance here is how recent trends in lighting design have become embedded within these developmental directives.

People entering Bjørvika from the train station, or coming from *Grønland*, a district across the tracks to the north of the OOH, will encounter an array of designed lighting as part of these new urban transition zones that 'function as the glue and the catalysator between the programs and as the new cultural, social and commercial meeting places for Oslo' (SLA 2020)[3]. These spaces are somewhere between public and private spaces, appearing as tastefully embellished mobility lines. Arriving in the foyer without a ticket to a show, individuals are drawn to document the aspects of design that catch the eye. Through their mobile devices or cameras, the foyer becomes just another stop, another interesting light play, that comprises their Oslo journey experience. In this sense, as a nodal transition space, the foyer echoes 'a certain kind of "publicness" in a privatised space' (Allen 2006, 442), adding commercial elements like a shop, restaurant, and café to the experience of art, architecture, and cutting-edge lighting.

Amin and Thrift have written extensively about the affective qualities of infrastructure and its mediation of our experience (2017, 51). Light as part of infrastructure has generally been utilised in connection with visibility and safety concerns (Ebbensgaard 2019, 95), but here we see that light and its related installations are part of a staging of the urban that seeks to improve the environment through feel as well as function. Moreover, in contrast to the use of specific lamp designs that, like urban furniture, help brand a city, the type of lighting in Bjørvika only becomes visible as infrastructure through its relationship with other materials, and as part of a transitional landscape. Other cutting-edge architecture and innovative designs combine with unique stonework,

Figure 6.7 *Akrobaten* pedestrian bridge with the *Barcode* development behind. Photograph by Bjørn Joachimsen.

horticultural arrangements, playgrounds, modern plazas, and bridges that use light to dematerialise their form.

John Allen writes that the idea of disorder promoted by writers like Richard Sennett tends to be designed out of development by affording specific social practices that in turn exclude others (2006, 243). In other words, difference becomes deleted. Whilst I concur, I suggest that here a notion of disorder, instead of being designed out of Bjørvika, has in fact been designed into the public spaces of the OOH and its surrounds, but that this does not in fact result in greater levels of participation. I will take a small detour here, in order to give an illustrative example using one of Bjørvika's bridges that uses intentionally smashed reinforced glass. Its design recalls the so-called *broken windows theory* that argued anti-social behaviour increases when visible traces of vandalism are evident. The solution it proposed was the quick and consistent reparation of any broken materials in order to create an atmosphere of order (Kelling & Wilson 1982). In the bridge's expression, the disorder of vandalism is built into the finished result making it impossible to break and unnecessary to repair. Jørgensen and Borch (2018, 14) argue that reading the broken window theory through Sloterdijk's conception of immunological architecture suggests that membranes, in this case, the glass, are essential to preserving the immunological value of a space, because they offer a feeling of protection from the outside. The membrane of the broken window bridge has already been compromised, yet we know of course, that the bridge is intact. At night the glass is illuminated, highlighting its novel material presentation whilst gesturing at the disruption of elements of its urban space. Relatedly, artist Olafur Eliasson has noted how contemporary museums have to make their ideology accessible to visitors, showing the conditions for the users' experiential encounters (Eliasson & Irwin 2008, 55). I suggest something similar is being conducted on the bridge; the shattered glass turns disorder in on itself, the boundaries of atmosphere become prised open, and we see the potential to transcend the hegemony of the *traditional* institution. Yet, this withdrawal is temporary, as Philippopoulos-Mihalopoulos warns us, even when we manage to escape from atmosphere we end up 'affectively directed, into another atmosphere. There is no escaping the continuous rupture into atmospheres' (2016, 164).

The bridge becomes another example of how light and materials, as part of Oslo's greater urban development, allude to the opening of borders and the potential emergence of atmospheres that end in a homogeneity of experience. It is in the repeated openings between membranes that Sloterdijk's notion of 'foam' gives us an understanding of our modern habitat as a 'mass container' (in Funcke 2005, 4), one which subsumes multiple different spheres into one encompassing atmosphere. A designed disorder is developed through the use of elements, as aesthetically contrived spheres where affects can be transmitted contagiously among the other foam cells (Borch 2008, 564). The zones that link Bjørvika to the rest of the Fjord City were designed as the 'glue' and 'catalyser' that would define each space as a unique area yet part of a cohesive whole (see SLA 2020). These transition spaces between new developments stitch Bjørvika together through an assortment of highlights that are varied yet similar in the ideals they communicate.

Images from Instagram reveal how people bring their experiences from the outside with them into the foyer, as one cutting-edge installation follows another. Their mode of participation is to playfully interact with this staging of place, seeking out the aesthetic openings presented to them. In this way, light combined with movement forms a mode of participation that reveals the OOH to be part of a greater conditioned environment, contained within a 'glasshouse' continuum that holds Bjørvika and its affects.

From bubbles to foam

It should be clear by now that a tacit driver behind the OOH's conceptualisation was grounded in the ideals of access, mobility, and democratic space. A notion of the Opera as an elitist institution becomes reworked through new art forms and structures that align high culture with social ideals. Interpreted through Sloterdijk's writing, we can read the original Opera house as a type of 'immunological architecture' (2016, 499), one that secured its hegemonic status by ensuring no type of participation could exist outside of the *art*. Here, the user was a spectator and nothing more. Contrastingly, the present OOH elicits openness and potential (see Payne-Frank & Schwabe 2022), refuting the perception of an elitist cultural institution. As I have shown, light as atmospheric element plays an important role in breaking down the *feeling* of a static and bounded institution, allowing its spatial borders to be perceived as porous.

Building on the themes from the previous chapter on movement, breaking down the distinctions between inside and outside manifests a connection to individual nature experiences, specifically local understandings of Nordic light. We can see how Nordic light becomes a conceptual tool for architectural design, where in the forms of *dark* and *liquid* light its rhythms are activated. In turn, an attunement to Nordic light reveals the cultural resonance it has with local people, and to a lesser extent, the foreign visitors who understand something of the Nordic. Such understandings are elicited by the social characteristics of light, for example, when the OOH becomes a lantern, reversing the process of sucking in the natural light from the fjord and releasing Nordic light back to the city. In its lantern state, the OOH was found to carry social tones, communicating a homely and cosy shelter. Here, light's relationship with materials, specifically wood, is essential to the warmth of light and its cultural emphasis.

Returning to the OOH's foyer we find the natural rhythms of the outside, staged through material filtrations and state-of-the-art lighting fixtures, the transitions between natural and artificial light. The passages between states of light and dark, natural, and artificial light echo the movement of people, creating a vibrant inside that runs through the OOH and greater Bjørvika area. Walking across the *akrobaten* bridge from Bjørvika to the Grønland district, there is a point on the other side when an entirely different space is entered, one that lacks the invitation to participate. It is here that the forms of dematerialising light and architectural installations cease, at the edges of city zones dominated by other aesthetic design strategies. What this reveals is the way light dissolves the spatial boundaries of the OOH and extends its atmospheric borders.

Throughout this chapter, I have highlighted certain transition points installed by Snøhetta as a means of engaging visitors by creating 'quizzical moments' that smooth their journey through different spatial zones. Thinking about the way light stimulates movement both physically and in relation to the emotive, I apply a notion of *air conditioning*, as a means of describing the way a sphere becomes 'moved, coloured or lit' (Philippopoulos-Mihalopoulos 2016, 152). When visitors arrive in the OOH's foyer, they have already traversed an array of new infrastructures, lighting fixtures, and designs, as part of an entirely new city zone. What air conditioning points to is that the OOH's atmospheric toning of space does not end at the edges of the foyer or its roof, but as I have described, at the edges of the Bjørvika development. Recent experiences are carried into the OOH, allowing individual spheres to form. What appear to be individual acts of participation, are, in fact, co-isolated spheres. As light erodes the membranes between inside and outside, individual participation becomes part of the 'foam' of experience, held within a greater atmospheric continuum.

What is shown to be at stake is the question of how design is able to nurture different forms of participation through elements such as light that can claim to open individual agency, whilst enforcing specific cultural meanings, as well as coercively tone whole areas with homogenous forms of experience.

Notes

1 The camera obscura, 'a dark chamber, with a hole punctured on one side, through which light streamed, forming an image on the back wall' (Otter 2008, 26).

2 'The one instrument that was ubiquitous in romanticism and the one that best served the needs and purposes of the romantic quest for the harmony of nature was the Aeolian harp. [...] The Aeolian harp is a stringed instrument played by the wind. The most common form is a rectangular closed box about three feet long, six inches wide and three inches deep'. For the Romantics, 'the Aeolian harp was superior to other instruments because its music was unpredictable and because it was played, not by man, but by the breath of Nature herself' (Hankins & Silverman 1999, 87).

3 Designed by Danish landscape architects SLA, 'the challenge was to create a masterplan and landscape design that could contain and create synergy between the many programs and typologies of the brief: A new state opera, two art museums, a university library, research institutions, retail areas, a new commercial and financial hub and many new mixed residential units' (SLA 2020).

7 Art participation

Sitting in the foyer and observing the space as people came and went, I became aware of subtle changes occurring in the periphery of my vision. At first, the gentle alterations in lighting conditions were difficult to locate, appearing less apparent when I turned my attention directly to them. The foyer presents light from various sources, which meant it was difficult to ascertain how the changes I was perceiving were being produced. Over time, I realised the variation in light was largely caused by *The other wall* and its almost imperceivably slow pulse between green and white light. Moreover, *The wall's* interaction with the everchanging natural light entering from the outside combined to stage potentially unlimited variations between light and materiality. Shrouded in these subtle fluctuations, *The wall* is an example of contemporary art's tendency to exchange narrative and form with 'movement, light and the gaze' (Rancière 2019a, xi). In other words, the prioritisation of spatial experience over what a work means, via the staging of atmosphere. Embedded within architecture, art becomes complicit in a form of spatial consumption, encouraging individuals to respond creatively to the space they are now immersed in.

Following on from the discussion of light, here I take up the theme of art as a mode of participation. The OOH contains several notable artworks commissioned by Norway's public arts body (KORO), but I am drawn to Olafur Eliasson's installation *The other wall* (2004–2008), not least because the artist explicitly engages with themes around atmosphere, individual agency, and audience participation which relates to the way people engage with the OOH as a form of 'dissensus' (Rancière 2009, 2010) and the activation of spectatorship and politicised aesthetic practice (Bishop 2005, 102). In order to develop a discussion around art's political potential, I keep Rancière's ideas in mind. His text, 'the emancipated spectator' (2011), speaks to the tensions between creative works and how people participate with them through social media practices, how they might open a space between 'conventional categories of participation and identification' (McCarthy & Wright 2015, 148), generating the possibility for politics to occur. In this way, *The other wall* exemplifies the OOH's entanglement in the 'aesthetic sphere', both removed from politics as a space for independent experience, yet always political with its promise of a better future (Bishop 2012, 27).

DOI: 10.4324/9781032668802-7

Figure 7.1 Detail of *The wall's* geometric structure. Photograph by the author.

I, therefore, use Rancière as a means of decoding *The wall's* role in shaping interactions and understandings between the viewer, the work, and the artist, adding a valuable third thing to the relationship of elements that comprise atmospheric experience. Part of my analysis explores the ubiquitous use of new media in art/architecture spaces. Here, Rancière's ideas imply that digital media shapes new distributions of the sensible, disrupting the relations between art object, artist, and viewer. However, it would be cursory to dissolve 'platform urbanism' (Barns 2020) into the notion of a distributed sensible, without taking seriously the specific communications between individuals across different spatial and temporal dimensions. Therefore, an underlying aim of this chapter is to shed light on what Degen and Rose refer to as 'the new urban aesthetic' (2022), and the relationship between new digital media and experience, in the context of art, and architecture.

Following the trajectories of *The other wall* develops an understanding of the way it shapes perceptible 'surfaces', manifesting certain understandings by partitioning the sensible (Rancière 2019b). If for Rancière the art object is a 'third thing' (Bishop 2011, 6) separated from both viewer and artist, then social media platforms form a fourth, linking the art space with individuals through communicative flows. User practices like capturing and posting author ever new *objects* that circulate as online content. In this way, *The wall's* role in the greater OOH development becomes scaled up, conditioning the very air with

Figure 7.2 In darkness *The wall's* light begins to interact with the foyer's materials and form. Photograph by the author.

transmissions between senders and receivers. Interpreting these transmissions as they resonate with different *surfaces*, I investigate how *The wall* (re)constitutes themes around ecological thinking, collective processes, and individual agency, but, also that its hidden structures and integration with architecture problematise its status as art.

First Encounter with *The other wall*

The wall itself is made of 340 m^2 of white tiling, installed in front of three large free-standing volumes. The installation uses two light sources, white fixtures set into the marble floor shining upwards, and green and white lights hidden in the inside of *The other wall's* panels. A control unit slowly fades the two light sources in and out, so when one is increased the other is decreased, the intention being that 'operagoers will thus perceive changing lights and shadows that influence their perception of the panels as well as the foyer's spatial properties' (Eliasson 2021d).

From this basic deconstruction of *The wall's* material arrangement, we can detect the tone of Eliasson's artistic thinking. His work, in conjunction with art and architecture, challenges assumptions of static built space by inserting a dynamic spatial quality. The artist's use of a material surface upon which fluctuating transitions play out implicates the surrounding area creating a dialogue between the work, space, and viewer. *The wall* thus anticipates a kind of co-production, engaging the viewer as they look at, move around, and sense the work's subtle oscillations. As Eliasson explains, the art object becomes a 'meeting-up of trajectories. Only at this intersection does the artwork become what it is' (2016, 1).

The wall's dynamic presence utilises natural and artificial light, shifting between architecture and art, the material and immaterial, in barely perceivable rhythmic waves. *The wall* transitions between different states, which as we have also discussed in relation to natural light, produce a multitude of moments and a feeling of infinite possibility. Moreover, its fluctuations suggest an intent to engage, emancipating the spectator from static representations and attuning the user to notions of democratic space and openness, thus following key narratives in the OOH's design and conception (see Payne-Frank & Schwabe 2022). In this sense, *The wall* proposes a departure from the traditional institution and its elitist art forms, actioning '*dissensus*', Rancière's term for the political act of disagreement necessary for a redistribution of the sensible (2009, 26). Crucially, art here is not held to be an autonomous creative act but part of a fabric of meaning, a 'surrounds' that draws upon processes of communication in order to shape cultural understandings (Turner 2013, 3). Such art implies a reversal of terms between the artist and viewer, where the viewer is empowered via subtle openings designed into the structure of the work. The paradoxical nature of a designed open lies at the heart of discussions around arts participatory credential, wherein Rancière describes the opposition in spectatorship as an enforcement of a distance between what the artist presents and what the spectator can learn. This distance results in 'stultification' for the spectator, which is overcome only when they are able to 'make it their own story' (2011, 22). By understanding the goal of liberating the viewer against the hegemony of what can be 'seen, thought, and done' (Rancière 2011, 47), *The wall* provides an opportunity to explore art as an invitation to participate, asking what kind of participation we are in fact talking about? (Bishop 2005, 6).

Democratic surfaces

Malcolm Miles identifies two general types of public art in the city: public art that inscribes the city, and art that has social intent (Miles 2015, 55). The OOH's use of art has strong social foundations built into the regulations of development whilst Eliasson's own iconic status as an artist adds a dimension of place branding. This complicates reading *The wall* as part of a move from 'monumentalism towards art forms intended to surprise, spark the imagination, or narrate hidden vernaculars' (Amin 2008, 17), because part of its 'hidden vernacular', relates to the mythology of the artist creator. Looking at public art's role in development projects can be a means of probing 'the argument of the eye', and investigating how ways of seeing relate to issues of power, knowledge, and individual practices (Hawkins 2013, 195). In keeping with this book's general atmospheric perspective, art like *The wall* forms a surface whereupon certain meanings and public expectations are played out between public culture and public space. However, through its integration with the OOH's architecture, *The wall* takes, as well as shapes, the architectural conditions of the space, which as stated previously, becomes interpreted as generally positive, open, and democratic.

The term, 'Installation art' is today applied to any arrangement of objects in a space which foreground the viewer as an embodied presence (Bishop 2005, 6). Users are 'installed as a body' (Hawkins 2013, 196), activated in ways that traditional pictorial art does not afford. I have discussed the tendency to dematerialise space in relation to architectural practice, and relatedly, *The wall* exemplifies a trend in installation art that seeks to turn entire spaces into immersive environments. As with architecture, the goal is to disrupt the artist–work–viewer relationship by emphasising experience as the work (Bishop 2005, 10). Art theorist Rosalind Krauss noted that sculpture, since the 1960s, had moved beyond the walls of the gallery into an 'expanded field' where the alignment of art and architecture challenged the purity of the art object (1979). More recently, 'relational aesthetics' have rejected the art object altogether to nurture social interactions (Bourriaud 2002, 14). By focusing on the relational, the artist becomes a curator, reduced to facilitating the collaboration of others (Foster 2006, 192).

What the varying examples of installations art share is a concern for audience activation, 'positioned against its mythic counterpart, passive spectatorial consumption' (Bishop 2011, 2). Rancière makes a similar observation, noting modern theatres internal opposition between truth and the exteriorising capacity of spectacle (2011, 5). Some critics worry that contemporary installation art slips too easily into 'an already pervasive culture of special effects and faux phenomenologies' (Foster 2013, 96), culminating in a relationship that does not so much give rise to 'the birth of the reader', as it does the 'befuddlement of the viewer' (ibid., 2006, 194). On the other hand, when art makes a union with architecture it operates at the edge of disciplines, 'exploding' distinctions between them which can give rise to new forms and potentials (Rendell 2006, 43). Whether the relationship between art and architecture eventuates implosion or explosion, art located across multiple disciplines, what Rancière terms the 'aesthetic regime' (2019b), requires new understandings, particularly in light of its insistence upon the equality of all subject matter (Tanke 2010, 10). *The other wall*, therefore, exemplifies an artwork that cannot be read as an autonomous artwork, but one conjoined through the equality of individual interpretation and its potential to allow individuals to walk their own *path*. Here we might recall how visitor's movements on the roof were shown to be both liberated and discreetly constrained by the design and use of materials. Related to the concern with individual participation, *The wall's* ongoing dialogue with natural light manifests themes relating to time, climate, and human/world relations. These new ecological themes often relate to an awareness of the environment in decline, temporal and interactive processes, and that our life is itself part of a dynamic and complex network of entities (Corner & Hirsch 2014, 259). I advance that this new ecological paradigm is subtly made manifest through the intricate material and immaterial surfaces of *The wall* in ways that will unfold in the coming pages.

Speaking to an architect from Snøhetta, they reflected upon how working with artists provided an opportunity for new ideas, and to experiment with the way materials react to light changes. I have previously discussed how the architects interpret *The wall* and its cold characteristics as a sensorial opposition to the warmth of the foyer's wooden wave wall. As with their own work, Snøhetta do not attach specific meaning to *The wall*, only pointing to its contrasting characteristics. Another of Snøhetta's designers, involved in the process of selecting the art, told me that they too detected a tension between the two walls, explaining: 'the art is integrated but also not, we didn't think about the ice motif which is what people now say about it'. The visitors I spoke with in the foyer also expressed a common recognition of natural themes in *The wall*, most frequently ice, icebergs, glaciers, and ice flows. An essential part of designing works open to individual interpretation is to avoid definitive symbolic meanings, so it is interesting that such themes have now taken hold. I once overheard a tour guide describing *The wall* to her group as 'reminiscent of ice crystals', whilst Kroksnes writes about Eliasson having transferred, 'the glacier theme of the opera house's exterior to the inside of the building' (2011, 120). With these narratives circulating in and around the house, it becomes difficult to discern between the works inspiration, its intentions, and the perceptions of individuals. Yet, despite the dominant interpretation of natural themes, neither the architect nor the artist commits to any fixed reading of the work, preferring to hold meaning open for the individual. In the absence of any affirmation of the work's intent, the agents of dissemination are left to fill in the gaps, adding fixity to a work designed to be ambiguous.

Looking at *The wall's* material form, we find similar examples in several of Eliasson's other installations. Most notable is his 'music wall' (2006) which presents a near identical geometric structure, albeit using platinum-glazed tiles that interact with the semi-reflective surfaces of Sønderborg's music hall (Eliasson 2021e). In fact, Eliasson has developed a series of façade walls consisting of complex geometric arrangements, using colour and light to create a dialogue with their environments. The artist has also collaborated on several architectural projects in the last decades, to list a few: the Copenhagen opera house, ARoS in Aarhus, Reykjavik's Harpa conference centre, as well as a complete architectural work in Vejle, Denmark (see Fjordenhus 2020). Informants were often aware of Eliasson's contribution to these other urban projects, whilst one of Snøhetta's designers admitted that, although *The other wall* was a site-specific piece, it was hard to read the work in isolation. Eliasson's profile in Nordic lands is now matched by his global recognition, and his Danish Icelandic heritage easily translates to western minds who discern the modest gesturing and democratic manner of his work (Cooke 2015).

Whilst participatory art shies away from the symbolic, much of Eliasson's work begins with heavy natural and environmental themes such as melting blocks of glacial ice (Tate ETC 2019), or riverbeds running through galleries (see Kortbek et al. 2017). These themes are often approached through a steady obsession with light, colour, and reflection inspired by natural phenomena (Frichot 2008, 32). Inspired by environmental wonders, Eliasson turns natural spectacles into site-specific spaces, inseparable from their surroundings. Most famously, *The weather project* (2003) attracted huge crowds to the Tate Modern's Turbine Hall, wherein visitors bathed in the yellow light of an ersatz apocalyptic sun. I visited the *sun* myself and recall how would people lie down to gaze up at a mirrored ceiling and a peculiar hazy vision of themselves. The work produced a feeling of disorientation, of being both within and outside the work simultaneously.

Having these works in mind, as some visitors to the OOH do, it is easy to read *The wall* as a further exploration of immersivity through ersatz weather and atmosphere. Several interviewees mentioned the sun in relation to Eliasson's wall. Notably, these prior experiences need not be drawn from actual situated encounters but can also arrive via cultural mediation, as demonstrated by one employee who admitted:

> I knew about the artist so maybe I had a preconception, and I read about the house before I worked here. It's like the sun in the Tate modern that made everybody lie on the floor, although I only saw it on the internet.

The employee identified *The other wall's* immersive qualities as similar to the sun, despite the discrepancy in scale, and although they had never witnessed the weather project live, they made a connection in the way the spaces were being atmospherically charged, how the spaces were 'no longer something perceived at a distance, but something within which one is enclosed' (Böhme 2013, 5).

However, as an integrated artwork *The other wall's* relationship with its surroundings is complex. In darkness it becomes a 'luminous object' (Böhme 2016, 195), a 'fundamentally aggressive' phenomenon of light (ibid., 196). I recall the creeping presence of *The wall* in full illumination, and how, contrary to its solid white state, in darkness people in the space are drawn to it but seldom touch. By observing *The wall* over time, using time-lapse photography and lengthy periods in the field, it becomes more than a contrived phenomenological encounter moving between art and architecture, and the immaterial and material. In this sense, the work can be read as a literal illustration of dissensus, presenting the viewer with an inexhaustible supply of unique encounters. The architecture as an institution is summoned through geometric form that is then obliterated as its solidity is dematerialised in an endless cycle. Eliasson's trick is to add a counter-point, a backlight that works against the illusion of solidity, changing *The other wall* into a 'quasi-object', an atmospheric thing extending into space (Böhme 2016, 183). Testimony to this is the way people engage with *The wall* when it is in its most solid material state, as an employee noted

Figure 7.3 Two young visitors framed by shadow lean against *The other wall* and have their picture taken. Photograph by the author.

explaining that: 'people don't just sit around, they interact, go and feel what it is'. Several other informants referred to *The other wall* in daylight as 'origami like', and visitors were seen touching its white tiles, haptically exploring its material qualities. In contrast, in its lamp-like state touching ceases, people migrate to its glow but don't get too close. In this way, the ebbs and flows between *The wall's* two states of being – a white patterned surface and a green/white lamp – engenders different forms of participation from visitors.

In his use of intricate lighting and lamps, Eliasson presents us with a series of conundrums, adding to the 'quizzical' feel of the space. We can see the lamp as an aesthetic object or we can see with the light of the lamp, but not both at the same time (Nørretranders & Eliasson 2015, 21). The transition from seeing to feeling reveals the shifting borders of the work and *The wall*'s relation to its surroundings. As one local put it: 'it makes you curious but not in an invasive way, it makes you physically curious to go over there, if it was just a wall then it wouldn't catch your attention'.

Allowing designs to be permeable and in dialogue with their surroundings means 'the weather will always be a part of the show' (Eliasson 2001, 10). Eliasson states that when architecture is seen as a static representation it absorbs the visitor's presence in a 'suspended narration of knowledge', a displacement where the weather outside makes no difference (ibid.). This chimes with Rancière's argument that an enforced distance results in the stultification of the spectator; foregrounding weather as part of spatial experience can be seen as a nod towards the intended emancipation of the individual.

Democratic surrounds: *The wall* as weather machine

Eliasson is not unique in his inclusion of the outside as a means of opening the artistic experience to more democratic forms. Fred Turner argues that the music of composer John Cage[1] has a central role in the formation of the 'democratic surround', a term that evokes Rancière's distribution of the sensible, pointing to the 'multi-image, multi-sound-source media environments' (2013, 3). According to Rettalack, 'what Cage discovered was that the more minimal and permeable the disciplined process or structure, the more it reveals about the world outside its perimeters' (2004, 178). For example, Cage's well-known composition 4'33", titled to describe the precise length of structured silence to be performed on piano, foregrounds sounds both inside and outside of the recital. There are parallels to Eliasson in how Cage thought about his music atmospherically and in relation to weather. His compositions contain 'semipermeable membranes that participate in dynamic, fluid exchanges with the elements that surround them: ambient sounds, ambient thoughts and emotions, cultural weather' (Retallack 2004, 191). As with *The other wall*, Cage's score assembles intensities and processes, becoming meteorological themselves (Herzogenrath 2009, 225). Both artists create a structure that gestures indeterminacy, allowing the weather to become more than a series of atmospheric effects and more a 'complex, dynamic, open, and thus unpredictable system of forces and intensities' (ibid.). Their work underlines the experience of nature always in emergence, and visitors are invited to realise their own part and explore in different ways as the borders of their experience are constantly being redrawn. Eliasson describes this form of engagement as a 'question of visualising our senses, making them evident through their instrumental representation: seeing yourself sensing' (Eliasson & Jensen 2000, 16). As with the mirrored ceiling of the weather project, the subtle ruptures *The other wall* makes in people's perception have the potential to make them alert, albeit momentarily, to how they are encountering the world. Thus, *The other wall* advances an emancipatory ideal, reflecting the belief that 'the artwork belongs as much to the person who meets it as it does to me. It belongs to you; it takes shape in your presence' (Eliasson 2016).

In 'the ecological thought' (2018), Timothy Morton reflects on the work of art as a whole made up of parts: materials, the context of the work, and the experiences of individuals. By acknowledging these external parts, it becomes 'obvious that there are so many more parts than there is a whole [which] means art and art appreciation won't stand still' (2018, 113). Morton draws a new ontology, making it impossible for an artwork not to hold multiple new experiences for the viewer. In a sense, Cage and Eliasson share this ontology excess, using the ineffable to push back against artistic control. Frameworks are constructed that allow new types of participation to play out through 'the reproduction of the dynamics of natural processes' (Herzogenrath 2009, 223). Presenting nature's 'manner of operation' (Retallack 2004, 205), *The other wall* opens a space for users to engage with the sights and sounds of the world (Turner 2013, 129). Yet, we should not assume that potential results in outcomes, and that *The wall*'s open structure brings about a redistribution of the sensible without further investigation.

Artificial light has typically been used as an architectural medium for the illumination of its facades, producing chiaroscuro that models the building by etching shadows into its surface (Plummer 1987, 29). Using light in this way enhances a building's solidity and draws boundaries, giving it a feeling of weight and permanence. Consequently, this lighting technique is often reserved for state institutions that want to enforce a position of hegemony, reminding us that lighting design is never benign but 'an active and affective agent, responsible for withholding or disclosing' (Barnaby 2017, 120). *The other wall*'s use of up-lighting echoes the techniques of traditional architectural lighting and is a key part of its structured opposition to the hegemony of spectacle and critique of the elitist institution, presenting us with a surface that dissolves before our eyes, again, and again.

A local visitor familiar with the OOH noted how a nature theme ran through the entire building. The changes between surfaces and light that had caught their attention outside were related to the dynamic light play of *The other wall*. And despite their recognition of the artificiality of the architecture, they explained: 'the art is like nature, and I think the way it interacts with people is warm and human'. As I have discussed in Chapter 4, the characteristics of materials carry social characteristics that in the foyer are placed in deliberate opposition; cold and warm, inside, and outside. Therefore, the feeling of harmony and balance detected by a majority of individuals I spoke to can be said to derive from the way the aspects of the building commonly present visitors with a lack of fixity. This, in turn, becomes understood as a progressive self-reflection from the side of the artist, architecture, and the institution.

Malcolm Miles writes that art generally reflects the shared cultural values of a society and that the conditions it creates are, therefore, more than the context for a work, but leave traces that have their own agency (2014, 11). Working within the frames of democratic space and the 'radical' institution (see Bishop 2013), *The other wall* renders natural themes, bringing a notion of the outside into the foyer space. Viewer participation becomes aligned with new ways of perceiving art and architecture, an 'ecological thought' that takes into account the complex networks of living things and flows of energy. Reflected in *The wall*'s more than material surface is a challenge to our current anthropocentrism, 'the world is teeming. Anything could happen' (Cage in Morton 2018, 54). In this way, *The wall* shapes the sensible, foregrounding an ecological sensibility that extends beyond the human social to include notions of environment, light, and weather as part of a new vanguard for future change.

Creative kitchens: process as art

'Process', writes Claire Bishop, has become a product (2012, 19). For art, attention to process highlights its potential as a social tool, disrupting 'the mechanisms of domination to turn the spectator into a conscious agent of world transformation' (Rancière 2009, 45). Eliasson apparently shares such concerns when he writes,

> in the cultural sector I think we make the mistake of claiming to be this moral compass for society, but often fall into the trap of being arrogant and elitist. We are caught in an echo chamber, a closed circle.
>
> (Tate ETC 2019)

The institution is held up as a standard bearer for a hegemonic culture that must be challenged in order to liberate artists and viewers. Yet, his work continues to collaborate with the world's leading art bodies, as if attempting to rupture the unjust and unacceptable relationship between institution, art, and viewer from within. As I have argued, *The other wall* attempts a move from the traditional culture institution to the sphere of human relations, following the imperatives of participatory art. And, I have highlighted how *The other wall* presents itself as 'an ongoing project without closure' (Larsen 2006, 173), using light and materials to shape an immersive space that blurs the distinction between production and perception. Its surfaces carry echoes of the grand process of shifting natural light and the movement of the sun. Here, I would like to exemplify this foregrounding of process in another way, one that refers to the social, or at least the relationship humans have with the world.

I take as an example one element of Eliasson's Tate Modern's retrospective from 2019, where visitors were given the opportunity to consume 'local organic food', prepared by the kitchen team of his studio in Berlin. The *work* is said to emphasise the communal aspect of eating and sharing ideas whilst practising care for the environment (Eliasson 2021c). By aligning food preparation with art production, the focus on process is held up as an ethical and politically loaded choice. There is history here, for example, in the use of dogmatic instructions for the crafting of Dada poetry[2] that attacked the notion of individual artistic genius, as well as giving the individual a direction to follow 'a liberating life praxis' (Bürger 2014, 53).

Studio Olafur Eliasson (SOE) highlights its kitchen through its personal Instagram profile. Here *#soe_kitchen* engages followers in dialogues around the food they first prepare and then post. Placed alongside Eliasson's other art projects, *#soe_kitchen* emphasises the lack of autonomy of any one work. Meanings cross and circulate among the mass of collective pieces, which in turn, accentuates a logic of the collective process. Consequently, with the notion of participating with the art, lofty ideals become bound to a plethora of individual lifestyle choices and expressions of identity. At *seo_kitchen*, posts about the staff food often refer to Eliasson's other work, as rainbow colours and geometric forms are echoed in the presentation of meals. Other images employ a coffee table aesthetic, presenting eye-catching images with underlining ecological themes, encouraging us to slow down and take note of the minutia around us. Fresh produce complete with stowaway caterpillars, uniquely shaped root vegetables, reinforce a message of novelty and difference.

However, like the vegetables, difference here appears to be cultivated with intent.

That Eliasson's kitchen is bestowed such importance, even to be part of an international art exhibition, divulges the extent to which self-critical antagonistic thinking has become embedded in the sensibility of contemporary art. *The other wall* enforces this antagonistic logic, rejecting works of individual genius whilst pursuing collaborations across multiple media. Emphasising a process of emergent meaning applies an 'institutional critique' (Palmer 2020, 319), alluding to the hegemony of top-down structures. Employees of the SOE are portrayed as liberated co-producers who believe in what they are doing, contributing to a more just and equitable way of life for people and the environment. Even as *The wall*'s surfaces entice us, it's light drawing us in, the process of creation becomes the essence of the work, adopting a stance 'that knows we have only just begun, like someone waking up from a dream' (Morton 2010, 98). Reiterating the ideals of the OOH, it points the viewer towards a better future, acknowledging the flaws of the past. In this way, *The wall* presents a surface that speaks to the greater intentions of Oslo city planners, as a local expressed: 'now the city is going back to the people'. *The wall* becomes part of an expression of the democratic reclamation of public space, that demands new ways of seeing and doing things.

In its collaboration with the OOH, *The wall* proposes a new radical point of departure, antagonising against unequal and undemocratic ways of building society, one that has been shown as harmful to ourselves and the planet. Thus, it speaks to the potential of a radical new future by highlighting a recognition of previous mistakes. As Eliasson himself explains, 'where once we used to look to history for inspiration, we are now navigating with the future' (Eliasson 2014). On the other hand, arts relation to architecture aligns it to a variety of stakeholders and commercial interests. As Weiss writes, the 'integration of collectivism and entrepreneurialism in contemporary architectural culture mirrors the reigning business model of "group genius" which harnesses creativity, collaboration, and improvisation for the sake of financial prosperity' (2020, 237). Such a critique disrupts clear understandings of art's value as a cultural and social good and requires our further attention, particularly as the pervasiveness of the 'group genius' model is such that its practices often remain unquestioned (ibid.).

The wall through social media

We have revealed how process is emphasised by the artist and his collective, gathering ecological sensibilities that point to new ways of creating and being. Here, I look at how people participate with *The wall* through digital devices, 'photographing it, commenting on it, sharing and posting the results on social media for friends and followers to see' (Palmer 2020, 318).

Digital media platforms have turned the viewer into 'an expert in visual choreography and manipulation' (Degen & Rose 2022, 142), thus supporting the mantra of participatory art that 'everyone is a producer' (Bishop 2011, 2). Users create a considerable amount of data, available online, that I take to be part of an 'interpretive grid' which co-constitutes urban experiences (Degen & Rose 2022, 130), and implies that participation with *The wall* occurs at different temporal and spatial dimensions. Social media allows the users of urban space access to 'a world where everyone can air their views to everyone' (Bishop 2011, 6); therefore, the art object becomes a part of a greater experience, a 'datascape' (Beiguelman 2020, 37) where people creatively document their journeys without discrimination. Plates of food, views of the city, encounters with nonhumans, quirky body poses, reflections, and light play are all presented to us against a backdrop of art and architecture.

Digital platforms provide us a glimpse of this world spread across time, allowing us to enter the 'visual orbit' of people's social lives (Palmer 2020, 318). However, this is a curated world where images are selected with purpose. Moreover, as with architecture, social media is now a key component of the 'promotional machinery of the artworld' (Palmer 2020 316), giving individuals the dual role of creator-consumer. This digital affirmation of what Alvin Toffler famously termed the 'prosumer' (1980) engages users to spontaneously capture, edit, and upload. Captions can be added, and content thematised through hashtag categorisation. Experience becomes bracketed and categorised. For the viewer online, multiple viewings and reactions to the post occur, even as the content producer remains in situ. In this way, posts relating to *The wall* take us *backstage*, but always to the fragments that individuals and algorithms select for us to see. Played out in public, these hidden social relations unfold between the spectator, the work, and an unknown audience.

The relationship between a post's non-situated audience and *The wall* is difficult to quantify because it falls into unknowable temporal dimensions. However, as art and architecture utilise social media for promotional purposes, it is germane to consider the aesthetic work they do and how they are shaped and practised. As discussed, *The other wall* presents a dynamic dematerialisation of its elements, atmospherically using light and materials to aestheticise the foyer space. Consequently, the documentational tourist photograph is now superseded by the pursuit of unique images that take as their subject the immersive qualities of art and its enticing surface effects.

Urry and Larsen (2011, 18) note that although immersive 'hyper-real' spaces 'condense sight into the most immediate and visible aspects of the scene', different types of gazes emerge. Paying attention to where people were taking pictures in the foyer showed a diversity of images. From my observations, individuals would capture images of place, people, and self, usually in that order. They would often begin by taking a general image of foyer space from a distance before moving closer to explore *The wall*, finding angles and unusual compositions. Many ended their encounter with a portrait or selfie, whilst others would move directly to document their own presence, showing little interest in the foyer space. Overall, there was a noticeable attraction to *The other wall*, particularly when it was illuminated at night. People apparently intuit its 'Instagramability' (Leaver et al 2020, 162), as an aesthetic quality that would enhance their own collection of images.

Some visitors arrive equipped with DSLRs and GoPro devices, but an overwhelming majority of individuals participate with the OOH through mobile phone technology. It should be stated that there is a fundamental difference between mobile and fixed-use photography. DSLR users would typically spend more time taking pictures, take more images, and review their images more closely in the field. Using a mobile device, as opposed to a camera with its single function, means people are often squeezing the activity of using the phone in between whatever else they are doing (Beiguelman 2020, 35). Taking images on a DSLR comes with an understanding that the editing of images will generally take place later, after the photographer has left the space. In contrast, mobile devices encourage a spontaneous flow of overlapping tasks and responses and the act of image-taking becomes just another task, no more essential than editing, cropping, and uploading. The upshot of this form of photography is that the image becomes less 'a stand-alone media object' and more 'a unit of communication', what Nathan Jurgensen describes as a 'social photo' (2020, 9). At the same time, this mode of participation destabilises the work of design, reducing it to the facilitation of experience where visitors are as important as the art object (Kozinets et al. 2017, 9). In the OOH's foyer, participating with *The wall* becomes part of the individual's process of a documented life, one which reduces experience to a 'logic of what can be digitally captured and shared' (Jurgenson 2020, 65).

Social media, therefore, has at least two noticeable impacts on the mode of art participation: its changes the practices of photography and the consumption of designed urban spaces, and it democratises the act of experience, giving the viewer an active role in its ongoing social production. Following Rancière's line of questioning about arts ability to rupture the 'policed order' of a distributed sensible, we should consider how *The wall* gives people possession of their own energies (2011, 6), as opposed to enforcing an alienating form of spectatorship. Reviewing my fieldnotes I recall observing people's engagement at various times: 'Two young women lie on the floor, spreading out on the marble shifting positions. They talk whilst one takes photographic portraits of her friend using Eliasson's wall as a backdrop. Upon entry, I see two women standing by *The other wall* taking pictures of each other. A mum takes a snap of two kids in front of the Eliasson. A man knocks the tiles of the Eliasson curiously. A young boy looks inside *The wall* to see the lights behind. A man gets his photo taken in front of the Eliasson by his partner. And again, but now with what I presume to be his daughter. Then the older girl takes a snap of the woman and man. They continue experimenting with different combinations between adults and kids'.

Whilst some people show curiosity about *The wall*, most appear content for it remain background to their experience. *The wall* becomes wallpaper, an infamous *selfie wall*. Art in a contemporary architectural context brings it into relation with tourism and photography, where 'the so-called "real" world became thought of as one spectacular exhibition' (Larsen 2004, 23). Mobile photography tapped into this process, speeding up the process of tourist consumption. Scrolling through Instagram at *#oslooperahouse*, Eliasson's wall appears to be just one of many vibrant urban images, 'a succession of high-coloured pictures continually gliding before the eye' (Batchen 1997, 93).

People engaging *The wall* with mobile phones not only capture but look, they use its screen as a window, 'to think about the world photographically but also to observe the photos of the world' (Cruz 2020, 328). The noted tendency in early tourist photography was for tourists to look for *the* picture, creating a sort of reverence around the object (Urry & Larsen 2011, 149) whilst travel was seen as 'a strategy for accumulating photographs' (Sontag 1979, 9). Relatedly, Roland Barthes observed how the photograph transforms its subject into museum objects (1993, 13); taking pictures was a mimetic practice reproducing the curation of the museum, art institution, or architect. Conversely, today's mobile images are taken as part of the individual's ever-growing collection of life experiences that denigrate everything else as background. Like the tourist's photo album, the social media platform is essential for archiving and presenting an individual's life story as they assume the role of creator and curator.

We have cautiously outlined the position that social media encourages a democratisation of the art experience and that artists like Eliasson attempt to liberate the viewer from hegemonic art forms. In fine art literature, the target of contention is often the gallery, the sterile 'white cube', whilst participatory art attempts to challenge this model (Bishop 2004, 51). Since the 1960s, a whole tradition of contemporary art practice has focused on the museum and gallery as a site for political provocation and intervention. In a sense, Instagram's interface echoes the space of the white cube, its square-bordered pictures displayed evenly against a white infinity. Art like *The wall* becomes 'shining baubles designed to attract a reluctant and easily distracted new audience, effortlessly photographed and just as easily forgotten' (MacDowall & Budge 2021, 3). The reader may draw lines to the discussion around whiteness in Chapter 4, wherein certain practices, and ways of being undergo a subtle process of surveillance. In Eliasson's interest in the emancipating act of 'seeing yourself seeing', there also lies a danger of self-censorship through enhanced self-awareness, by which I mean people's participation can become limited to what they deem as aesthetically desirable, based on the inescapability and promotion of conventions and practices of others. In fact, it can be said that digital platforms function through a regime of imaging practices, based on their architecture, filters, and algorithm (Beiguelman 2020, 40). It is therefore important to point out the tensions between the potential for the creator's innovation and the restrictions of the medium, especially as these restrictions tend to be concealed beneath the platform's slick usability and glossy aesthetic. As with the example of Dadaist poetry, the structures of production can be both liberating and restrictive, exemplified by the apparent similarity that many images of *The wall* divulge, combined with the continuous search for novelty that is a part of this concealment. The technology of digital imaging *improves* continually, upgrading the capacity for users to participate with and through the screen. Degen and Rose point out that urban development projects communicate the 'offer of specific future aesthetic experiences', which goes hand in hand with digital technologies and their association with futurity (2022, 13).

Moving through images posted by creators over time reveals a history of filtering and effects. Earlier images have another aesthetic, quite different from today's deep shadowy contrasts and saturated colours. Most notably, the vintage style filtering was prominent in the early 2010s, with its black border, square format, and washed-out colours. According to Jurgenson (2020, 5), the attraction of these analogue-style filters was bound to the perception of digital technology at the time. Seen as inferior to the tactile physicality of the photograph object, the goal was to show an authenticity of experience, connected to the real-life encounters of the individual. Authenticity remains essential to today's aesthetic but is approached in a way that embraces the technology now deeply embedded in our lives. *The wall's* shadowy high-contrast look translates easily into the current aesthetic. Doctored images increase hues and tones, further enhancing its appearance whilst its form becomes distorted by wide-angle effects that transform its lines into bulbous and sculptural interpretations. Digital technology delivers the ability to morph *The other wall* into *new* works, acting as an authoring tool for the creative user.

Figure 7.4 *The other wall* altered through photography to a saturated and sculptural form. Photograph by Bjørn Joachimsen.

As part of the 'aesthetic regime', *The wall* gathers art, media, and popular culture together with the forms of life (Rancière 2019b, 107), showing art participation to be related to a 'fusion of media and bodies, making clear how we see, think, speak, and feel through images' (Jurgenson 2020, 112). Furthermore, these images are not simply an expression of the aesthetic regime but bound to it through an ongoing dialogue. The inbuilt filtering systems of mobile phone cameras add a layer of effect to the scene, an illuminating example of which came through an informant, a local man well acquainted with *The wall*. He mentioned how he enjoyed the different colours that the art produced: 'sometimes yellows, sometimes blues' he explained. These colours are, in fact, not present in *The wall's* white and green lighting, but only through the mobile device's inaccurate rendering of light. For my informant, the distinction between the *reality* of *The wall* and its many representations had become inseparable, further evidencing how digital photos constitute collective meanings and imaginative responses.

Capturing outlandish colour and happenstance effects are therefore shown to be a part of the co-creative act of experiencing the art. Many posts present *The other wall* in a grossly exaggerated hue, either enhanced post capture or in-phone, as part of people's creative take on the work. The immersive nature of Eliasson's work encourages people to play with the available visual effects, recognising that activated spectatorship requires them to give the work an individual *touch*, to make it their own.

Many images include figures captured in novel ways within the spatial structures of the space, becoming literally engaged with the art through their proximity. Whilst many visitors use *The wall* as a backdrop for a portrait, others step back to capture the flow of strangers interacting with the art in quirky, visually compelling compositions. Kozinets et al. note how art can become a symbolic resource for people to shape narratives of self, 'art works and museum spaces become props, background material, and stages upon which individuals act out the experiences that give their identity its uniqueness and their life its meaning' (2017, 9). For both the selfie taker and the documentary-style photographer, *The other wall* presents a dynamic and immersive space that compels people to seek fixity amidst the dynamic emergence of effects, in order to claim the art as their own.

Larsen writes that a part of the development of early photography stemmed from a desire to fix the airy and transient nature of technology like the *Claude glass*[3], into permanent forms (2004, 21). The notion of seizing nature's landscapes, capturing forever a view that could be owned and removed from the local vicinity, aligned it with the painting of the time (Morrison-Low 2005). With the advent of photography, painting and the arts began to explore the capacity for expression and the feeling of place, leaving photography to the work of mimesis. Böhme notes that the first photographic images were of architecture due to the long exposure times that required large static subjects (2016, 148). Art and architecture's recent turn to the creation of immersive and atmospheric spaces is now reflected in some of today's architectural photography that

Figure 7.5 The silhouette of a figure with a cap and trench coat is mirrored against the vibrant colours and geometrical structure of *The other wall*. Photograph by Morten Andresen.

attempts to draw out the phenomenological experience of being situated in a specific place (see Baltzer 2004). Böhme adds that the art of architectural photography lies in its ability to make visible, that which 'cannot be visually perceived' (2016, 153), implying photography should pursue the atmospheric feeling of a place, an endeavour recognisable in some posts which engage with the immersive qualities of *The other wall* and its surroundings. The manner by which *The wall* shapes a dematerialisation of the art/architectural object is enhanced by a medium which is itself flexible and always moving towards reinvention. Moreover, 'the dematerialised art object turns out to be perfectly adapted to the semioticization of capital' (Steyerl 2013, 7), which is to say, the mode of participating with *The wall* is manifestly entwined with social media's modes of production and its compounding of creation and consumption.

It is tempting to make a distinction between architectural photography and the photography of social media, yet the utilisation of social media by art and architecture blurs any clear division. For example, *#operaen_* is curated by the OOH's marketing department and creates its own hi-end content, but also reposts content from Instagram users it regards as on point with their own messaging. Eliasson himself has expressed his desire for people to engage in a more spatialised haptic interaction, and for his art to encourage people to 'put their phone in their pocket' (Ingram 2019). Yet, *The wall* is one of a series of works that isn't just atmospheric and immersive, but highly *shareable*. It appears that the anticipated 'mass empowerment' of the user might in fact result in an 'endless stream of banal egos' (Bishop 2011, 6), where individuals consume spaces in similar ways. This begs the question of whether the relationship between art and social media supports a democratisation of culture, or does it simply follow architecture's fate of being reduced to the 'texture and appearances, of playfulness' (Urry & Larsen 2011, 120).

Selfies and mirror selfies

There is a helpful distinction to be made between the different types of images people post which can reveal something about their mode of participation. Even the ubiquitous selfie is worth looking at in greater detail. The selfie can be characterised as a simple framing of the self, with the intention of being seen by an audience online. Yet the performative aspects around taking selfies can be 'complex and multidimensional' (Kozinets et al. 2017, 1) and require some unpacking as they are produced through a variety of performative traits. Moreover, there is a difference between people taking portraits and taking selfies, which relates to the way the hand-held image conveys a greater sense of authenticity, something 'akin to looking in the mirror' (Jurgenson 2020, 55).

Similar to the practice of creating novel images of place, in the selfie the hand-held image foregrounds the author of the scene. Placed in front of a work of art, the image reveals 'intimate aspects of the fabrication of the social and cultural' (Shanks & Svabo 2013, 237), playing with the authenticity of the image that is a hallmark of social media aesthetics.

Visitors often move quickly through the foyer, pausing only to document the highlights of the space. Having done so, they typically document themselves within an aspect of the space they find most enticing which in the foyer tends to be *The other wall*. Perhaps aware of the banality of their actions, many individuals attempt innovation and adopt a playful approach. In one posted image, a young woman can be seen in front of *The wall*, but appears to be taking a selfie of herself, looking at her phone whilst having her portrait taken. 'Mirror selfies' are a recent phenomenon in social media that involves pointing the phone at a reflective mirror surface. Claims are, that in so doing, a layer of novelty is added by mimicking the conventional selfie, thus knowingly revealing its production. We find here another example of the way works are compelled to reveal their structure, enabling the user to realise their role as a spectator. Typically, corners of a mirror or the edges of a surrounding wall will be included as proof of the image's real frame. In this case, we know that the young woman is not the taker of the image, as there is no mirrored surface for the image to reflect her. In the caption, the author has added: 'bringing back the mirror selfie', a tongue-in-cheek reference to the related trend of faking the mirror selfie, and contriving an image that is, in fact, a staged portrait.

This playful fakery of the mirror selfie presents us with an image of authenticity that is knowingly contrived, revealing the construction of designed environments that, as Foster points out, is also a tactic used by architects to suggest transparency and participatory intentions (2013, 213). Likewise, Eliasson's work often tends to present the scaffolding of his deception, simultaneously showing us the trick and how it is done. As with the mirror selfie, its intention is not to deceive us but to communicate an alternative approach, ultimately channelling us back to the individuality and motivation of the author. Likewise, *The other wall* presents us with an experiential space that shows us the means by which we are being made to feel. This simultaneity of immersion and alienation, held within the spatial experience of *The wall*, is problematic in the sense that people must both 'consume culture and enjoy the visual affirmation of yourself as participating consumer' (Lavender 2012, 326).

Sloterdijk makes a similar observation in his conception of the conditioned glasshouse as a state of inauthenticity combined with the goal of happiness (2014, 170). In this way, *The wall* shapes an ambiguous interplay between the emancipation of the individual, and a deterministic nudging of their creative content via various designed surfaces and platforms from which their experience can take place. Many visitors meet this challenge and participate through mobile media, finding their way amidst the tensions of 'a self in the active process of being made and also passively shaped by the world' (Jurgenson 2020, 55).

***The wall*, dissensus, and a partitioning of the sensible**

Focusing on *The other wall*, I have explored art as a mode of participation with Rancière's writings in mind. In identifying *The wall* as a work that carries a critique of representative art and the traditional gallery institution, the question becomes how *The wall* as a public artwork closes the gap 'between those who act and those who look' (Rancière 2011, 19). In tracing *The wall*'s various surfaces, this chapter has made account of some of the specific ways meanings and practices become (re)constituted through relations between the art object, the viewer, and the artist. These relations play out within the accords of a certain distribution of the sensible that transcends the art object itself.

By thinking about *The other wall* as embedded in the 'aesthetic regime' (Rancière 2009), and not as an autonomous object, the art becomes tinged by other disciplines, their media, histories, and politics, as well as the embodied experience of individuals. Furthermore, Eliasson as an iconic artist figure is brought into the space through individual knowledge of his previous work and persona. These are the complex conditions from which *The wall*'s modes of participation arise. Its status as a piece of integrated architecture holds a similar tension as do the claims of the OOH as a 'social monument', combining the iconic with the everyday. Part of the work's success can, therefore, be realised as its resonance with the surrounding architecture, following similar themes and contributing to the greater gestalt of the OOH, and greater Fjord project.

The wall's presentation of light and form was found to manifest egalitarian and ecological perspectives, which now the ecological has been freed from the demands of an authentic nature (see Hörl & Burton 2017) speak to any number of aesthetic themes. In fact, questioning the authenticity of nature is at the heart of Eliasson's artistic contribution, stripping nature bare to reveal its atmosphere of phenomenological affects. However, such ideas require individuals to invest their time, because it is only through durational exposure that *The wall*'s rendering of diurnal rhythms and natural forces become apparent. For those visitors in a hurry, seizing the immediate play of light before them appeared to be satisfactory. People who become familiar with Eliasson's work through time, light, and material form note its contrast to the wave wall and an antagonistic stance against the traditional institution. Its continuous swing from white material surface to immersive installation suggests an intent to dissolve the static and representational, allowing individual experience to flourish. Likewise, it opposes elitist ideas around artistic genius and the autonomy of the art object, nudging visitors to move, interact, and become aware of their own modes of perception. In other words, to liberate themselves from 'stultifying' spectacle. And yet, all of this is played out against the cutting-edge architecture of the OOH, and the iconic status of the artist himself.

This ambiguity is compounded by the use of social media and mobile technology, a main aspect of visitor participation with the OOH. Through the digital frame, individual authors share new interpretations of the art that can be regarded as emancipating, or in turn, as the enforcement of an aesthetics of antagonism. Importantly, the use of social media is not reserved for individuals, but is used by the institution, architect, and artist to control as best they can, the dissemination of their own projects. This leads us, finally, to the way the designers foreground process as a means of communicating ethical and democratic ideals. The creative act is portrayed as a collective activity where the outcome of the work becomes of secondary importance. The notion of collective process, both in the studio and with the world, bestows *The wall* with equality and an ecological sensibility that appears as the outcome of new and democratic ways of doing things. In this sense, process becomes the work itself, connecting discourses around the production of culture, environmental concerns, and individual lifestyle choices. The viewer is invited to partake in this ongoing creative act, although as Claire Bishop points out, just what the viewer is supposed to take away from a work resistant to closure, and the experience of creativity is unclear (2004, 53). Here, we see how dissensus is introduced through *The wall*'s opposition to hegemonic narratives. However, as Hal Foster writes, in his rendering of nature as a mediated phenomenon, Eliasson also alienates the viewer, turning their participation into being a spectator of the spectacle (2013, 213).

We now understand that dissensus can be discussed in the way it affords viewers the potential to create new works through innovative angles, colours, light, and interactive embodied performances, but that this turns the art into an element of composition as the participant moves from viewer to co-author. The images people take were noted to be varied in the sense that portraits and selfies were approached in different ways, as people playfully responded to the visual intervention that *The wall* makes in the foyer. On the other hand, digital interfaces, filters, and algorithm have a structuring effect on the practice of creating images, it constitutes quite specific forms of creation which challenge the medium's democratic claims.

Art participation through mobile technology and *The wall* as a means of dissensus becomes shrouded in ambiguity when we realise that its scaffolding can also be seen as a form of 'police'. It poses the question as to whether the agency afforded individuals is, in fact, simply their participation in the consumption of culture (Lavender 2012, 326). As Hito Steyerl points out, the images such participation produces hold a number of tensions which 'express all the contradictions of the contemporary crowd: its opportunism, narcissism, desire for autonomy and creation, its inability to focus or make up its mind, its constant readiness for transgression and simultaneous submission' (2009). As we have seen, individuals can be physically present yet absent, whilst those elsewhere can attune to the atmospheres of a space. Ultimately, individuals 'seeing themselves seeing' at the OOH might only be the fleeting recognition of their own participation with *The other wall*, as they flick between disparate yet entangled worlds.

Notes

1 John Cage was a composer with an interest in the readymade nature of music and art. His approach was that everything can potentially be art. 'The only thing that isn't art is inattention'. By focusing on the pauses and breaks in structures, he attempted to reconfigure the audience's attention (Retallack 2004, 176).

2 Bürger refers to the work of Tristan Tzara, an example of which can be found here: https://www.writing.upenn.edu/~afilreis/88v/tzara.html

3 The Claude Glass was a portable viewing device made from a tinted convex mirror. 'Travelers would hold the glass in their hand so as to see a reversed image of the landscape positioned behind their back' (Batchen 1997, 73).

8 Conclusion

Exiting the social monument

Since its opening in 2008, the Oslo Opera House has become a new icon for the city, a 'house' at the heart of Oslo's waterfront development that combines culture-led, cutting-edge architecture with social ideals such as access, egalitarianism, and participation. In this book, I have followed a general curiosity about what it is that people do when they visit the OOH, how the *invitation* to participate is put into action and becomes played out.

In this context, I have highlighted the OOH's claim to be a 'social monument', bringing together two contrasting perspectives: the iconic and the experiential, before locating a series of similar tensions in the present-day Opera house. These oppositions are traced from political debates around its conception and purpose, the rhetoric of its design brief and the intentions of the architects, to the perceptions of visitors. Amidst such ambiguity, it was remarkable that visitor's general perception of the OOH amounted to a comprehensible and homogenous architectural experience, an observation that lies at the heart of this book because it raises questions about how diversity and difference is constituted through architecture and development.

The contemporary trend of transforming urban spaces into experience-based encounters means the question of difference is being addressed through design. As this book has shown, architects and artists are now engaged in creating frameworks that allow different types of experiences to flourish. Enter the notion of atmosphere and its potential to stage urban spaces with encounters that we feel before we comprehend. For many Norwegians, participation with the OOH begins even before they arrive, as a fabric of understanding related to their exposure to the project's long and public conception forms part of their atmospheric attunement, compelling them to experience the OOH's architecture in accordance with how things are *allowed* to appear. It is here that questions arise around how cities are built, why, and for whom.

In many ways, the OOH exemplifies how urban development increasingly recognises the value of social diversity. As Sennett (2006) suggests, 'equilibrium and integration, just like over-determination, can be destruction' for a city. Therefore, the question of how diversity is allowed to exist in the urban context is crucial. At the same time, there is increasing recognition of the

DOI: 10.4324/9781032668802-8

determining power of design in shaping experiences in spatial, virtual, and even political arenas.

As part of Oslo's extensive overhaul of the harbour front, the OOH invites participation in particular ways, one of which is its status as a Nordic project. As a social monument, the OOH speaks to the idea that architecture should make a social contribution, following a Scandinavian welfare-based view that puts urban development at the service of societal goals. Forerunners such as Jan Gehl have successfully spread the 'cities for people' concept, gaining global recognition for their human approach to architecture, a case in point being Gehl Architect's involvement in the planning and recommendations for Bjørvika and Oslo's Fjord development, of which the OOH is a part.

Throughout this book I have outlined how making spaces for people tends to end in an engineering of the experiential, foregrounding the importance of understanding how atmospheres become the backdrop to people's urban interactions. Exploring materiality, movement, light, and art as separate but overlapping modes of participation, I have investigated the way atmospheric design supports multiple simultaneous meanings. Most strikingly, the rendering of the 'mountain' as a globally understood symbol of the Nordic was made compatible with deeply held understandings of the Norwegian outdoors and a local sensibility to nature. In so doing, the OOH speaks to the local and global, avoiding the exclusivity of the vernacular, and the blatant iconicity of international city branding. These understandings resonate through atmospheric encounters with the roof of the OOH, and crucially, through the body as it moves amidst an environment that is artificial yet open to the weather and fjord. Furthermore, a general notion of interacting with the outdoors stimulates tacit national identifiers and socio-political ideals. Movement, light, and materials atmospherically charge a space where the local and global, the social and monumental are shown to co-exist.

Just as experience emerges on the roof, it is important to maintain that as a tax-funded building, it was necessary for the project to emphasise its social and political ideals. Therefore, the notion of a social monument is not only played out by visitors to the roof but was conceived in accordance with certain political ideals. In contrast to its contemporaries, like the Danish opera house, the OOH stimulates democratic space by enhancing a flow of accessibility to and from the city. However, being a node within a greater network of mobility and access has led the 'house' to become what locals jokingly refer to as, 'the most expensive public toilet in the world'. The idea that a cutting-edge building is reduced to such base functionality appeals to some locals and their *janteloven* sensibility, whilst for those on the way to the *Sørenga* beach, stopping off at the OOH is an indication of the building living up to its promise of being a functional Norwegian public space. Similarly, the different modes of transport and activity such as bikes, dogs, and electric scooters, whilst discouraged, are not actively prohibited. To do so would be to enforce more than rules and boundaries, and to contradict the social claims of the monument.

In addition to the everyday tolerance shown by the institution, I have exposed how atmospheric elements such as light, materiality, and movement contrive porous borders which connect the architecture to its adjacent surroundings and create perceptions of openness and potentiality. The way people are encouraged to move is crucial to a particular form of embodied engagement with the building, and there is a clear relation between the roof's sensuous qualities and its perception as a 'blank slate' where people intuited a freedom of movement. Despite the OOH's monumental and iconic status, its lack of signposting was interpreted, particularly by children and the less inhibited, as an opportunity to appropriate the space and *complete* the work. One particular informant drew a parallel to the feeling of moving in an abandoned industrial space, revealing a sensuous attunement to the OOH's rendering of potential. In contrast was the photographer described in Chapter 4 who held on to the image of the OOH as a white building out of place, even as he stood amidst the activity of people on the roof.

The OOH's 'white carpet' stages these forms of movement, both constituting and being constituted by people's participation. However, as discussed in relation to *whiteness*, it is a surface that elicits and deletes difference. Here, whiteness shapes atmospheres of potentiality and futurity yet also acts as a type of surveillance machine, monitoring individual participation. The white characteristics of marble underline the OOH's social credentials by highlighting the potential for movement, but as the architects Snøhetta concede, the marble is complicit in a design that also directs people along certain paths. This points to the fundamental question around the emancipation of the user, which at the OOH is often in tension with the coercive restrictions of design. As Claire Bishop points out, the distribution of the sensible opens a door to both the progressive and the reactionary (2012, 28).

In the analysis of Olafur Eliasson's installation *The other wall*, it was shown how *process* was a form of antagonism, foregrounding an oppositional stance to the conventions of the institution and the one-way communication between artist genius and passive receiver. Similarly, the costume workshops at the rear end of the OOH were notable in the way they display the art of skilled employees, both as a visual feature and as a means of puncturing the illusion of spectacle on the theatre stage. Interestingly, this trend of revealing the artifice was also found in images posted by individuals online as they played with notions of photographic representation, spatial consumption, and identity. Alongside these examples of transparency and antagonism built into the OOH's design, I have tried to develop a sense of how the affective qualities of the space also resonate with and constitute similar themes. In relation to this, many of the architectural features and artworks were discussed through their sublime qualities, for example, the atmospheric rendering of Nordic landscape. I have pointed to several examples of how the OOH encourages people to spend time in its public spaces, yet much of my analysis highlights the seductive qualities that tantalise the visitor through 'quizzical moments' and 'soft transitions'.

Relatedly, we have seen how the language of the architects and artists is grounded in the tensions between ways of building. For example, Snøhetta's emphasis on the perils of verticality, and the virtues of the horizontal that draws lines to public space, democratic values, and the virtues of 'building for people'. Notably, the antithesis, in this case verticality, is never far away, summoning a notion of closed hegemonic control and the elitist institution. This typifies many of the analytical discussions within this book that focus on the way the OOH and its art sets up a series of ambiguities and oppositions, just as the experiential bent of architecture implies sensuous interaction as well as its adversary, the domination of the eye and passive spectatorship. Such tensions are repeatedly manifested, as in the use of marble as a material that speaks to 'the archetype of the royal palace' (Böhme 2017, 57), as well as the Nordic mountain landscape, a material embedded in the renaissance history of art laid out for people to walk on and shape through their own participation. This use of a counterpoint points to the democratic drive behind the work, whilst the resulting ambiguity can itself be held up as proof of difference being played out.

What I identify as an antagonistic approach to the institution, elitist art, and modernist ways of building results in the reproduction of binary terms, for example, presenting the OOH as open because it is *felt* not to be closed. This allows the OOH to assert a series of relatable ideals, thus horizontality is by virtue democratic. These terms become accepted because they fit into the logic of the building's overall project, just as they are re-constituted atmospherically through light, movement, and materials. An interesting side note is that the OOH's neighbour, the Deichman library, also very much a social space, is a vertical structure. The difference being that its social is played out in the inside and relates to more clearly defined forms of participation. This discloses that the OOH's roof is as important as a means of communicating the invitation to participate, as it is as a facilitator of people's engagement.

We should, therefore, be wary of allowing meanings and associations to quietly slip into frame and become assumptions but attend to atmospheres and the work they do. I have explored this lacuna considering user experience and the design intentions of the OOH, blurring easy distinctions between experience and design as people's engagement with place is tainted with understandings forged elsewhere. Accounting for the way people feel in relation to their interactions with the design of space has been a main analytical theme of this book, explored through an atmospheric lens developed over time in the field.

With Böhme's theory as a starting point, atmospheres are essentially phenomenological encounters between subject and object, which, despite his acknowledgement of design and an 'aesthetic economy', lean towards the individual determination of experience. Through Rancière and Sloterdijk, I have attempted to agitate Böhme's conception, in relation to the invitation to participate, emphasising that the potential for emancipation does not mean it necessarily delivers. Instead, the distribution of the sensible becomes the complex constitution of meaning behind the OOH that I interpret as being entirely relevant to atmosphere in that it tones the way we perceive its space through all manner of events, and cultural references, material, and immaterial elements. It is this, 'autonomy of experience' (Rancière 2002, 136) that I have attempted to account for, exploring the way it co-constitutes modes of participation as it directs visitors in subtle and unconscious ways against the potential of their own meaningful subjective experience.

In recent times, Rancière's project has been to account for 'aisthesis' (2019), developing a language that can address the complexity of meaning and the regimes behind a work of art. His texts depart from the taste-based judgements of aesthetics, showing us how art objects are consumed emotionally. In atmospheric theory, much of the discussion around human experience takes place in the abstract, but by bringing together different theories of understanding atmosphere, this book brings emotional responses into dialogue with a background of distributed meaning, using empirical examples as a springboard. It uses *participation* as a term to distinguish between a general experience of space and the experience of space that involves an invitation to act. And, by recognising the OOH as a building charged with democratic intent, the 'house' is used as an example of how people engage with a growing number of contemporary urban spaces that imply a shift from bounded symbolic structures to open and autonomous experience for all.

The final analytical chapter of this book wonders why the individual *journey* is expressed and interpreted as essentially homogeneous. The similarity of atmospheric experience was explored specifically through light, showing that its particular use extends the perceived boundaries of the OOH. Drawing on Sloterdijk, I explain this smoothing of experience by identifying the OOH as part of a greater atmospheric continuum contained within a 'glasshouse'. This speculative argument is evidenced by how light, architecture, and movement condition the air over large swathes of Bjørvika's new development. Visitors journey from one dynamic installation to the next, following a textured but cohesively similar environment. This raises an important point as it reveals how the relationship between design and the individual reduces their participatory practices to a form of 'hypnotic relations': believing they are acting spontaneously but are, in fact, only reacting to the structures and suggestions presented to them (Borch 2010, 226).

For those architects and artists who venerate the shaping of urban experience over representational forms, meaningful experience arrives through 'an aspect of awareness' as the individual reflects on their experience as a way of gaining insight into themselves (Boswijk et al 2007, 20). There appears to be consensus between the writings of practitioners such as Pallasmaa, Zumthor, and Eliasson, in that they follow the general terms of Böhme's atmospheric theory and an understanding of individual difference as the inevitable result of atmospheric experience. This is significant because the assumption of individual agency through the experiential can be traced to the manifestation of, and understandings around, the OOH's built space. What we can conclude is that if design is a means of achieving the individual's emancipation from passivity, it can only do so through sleight of hand. This paradox is reiterated in different forms throughout the OOH, making its ambiguity a central concern of this book.

Realising how atmospheres tend towards the indistinct means attending to the determinate and indeterminate qualities of a space (see Anderson & Wylie 2009) as individuals discover various openings and closures. I understand this ambiguity as an inevitable outcome of large-scale, multifunctional building, yet the way designers shy away to leave the determination of meaning to others brings to mind the observation that exclusion can work, not only through physical barriers, doors, and gates, but through 'a logic of inclusion' (Allen 2006), and that some forms of public space might not be as *open* as they feel. Furthermore, the ambiguity of the OOH's enigmatic design produces an opacity that means they can be converted to spectacle (see Foster 2013, 99). Here, we can recall the way people participate through digital devices, attempting to make the space their own and freezing its otherwise dynamic characteristics.

During fieldwork, it became ever more noticeable how people's engagement with the site was connected to their use of a mobile device, making it altogether unclear whether the space was constituting new forms of participation, or returning individuals to the passive consumption of space. This opens further questions as to the nature of participation in relation to people's interaction with space via digital media. The invitation to participate through social media platforms and digital media, further questions the notion of atmosphere as grounded in individual experience, just as the ruptures it makes in space and time disrupt our understanding of situated participation. The relationship between photography, digital media, and iconic styles of architecture is an area for future study which this book does not take far enough. One reason for this is the fact that digital platforms have become more challenging to use as data. The compliance tech companies show researchers are minimal, copyright rules have been tightened, and even large publishers appear reluctant to reproduce content posted in the *public* domain. Such constraints only underline the importance of shedding light on digital media in relation to the production of space and place.

That digital media unsettles the concepts of atmosphere and participation is in keeping with another theme developed in this book, that of dissonance, both as a tool for researching atmospheres and as a generator of meaningful attunement to place. Whilst resonance speaks to the way design can smooth experience, some of the most empowering examples of atmospheric affect emerged through a dissonant rupture in atmosphere. In the chapters on materiality and movement, I showed how dissonance through 'glare' and 'tripping' held the potential for visitors to participate in embodied and imaginative ways. From a design perspective, such dissonance arrives in the unforeseeable gaps of staged space, rather than as a direct result of their cohesive assemblage. It is in the *surprises*, the ruptures in the staging of atmosphere, that meaningful acts of participation occurred. This is altogether different from assuming subjective experience to be the defining constituent of atmosphere because it forces the outside world to be taken seriously as a co-constituent. Weather, for example, produces sensuous engagements between visitors and the outdoors, and the meteorological becomes something no longer simply out there, but takes on a new life as it is framed and presented through the architecture and the senses. Furthermore, the experience of phenomena such as heavy rain can be democratising. Consider the huddles of people sheltering in the OOH's foyer, having been caught in a summer downpour. They laugh and shake themselves down, glance at their neighbour knowing they have experienced something in common.

This brings us to the observation that dissonance tends to be hidden from view through the spatial arrangements of architecture. In the chapter on materiality, it was found that certain *stains* were tolerated more than others. The yellowing of marble, for example, challenged the OOH's whiteness, the sculptural and culturally desirable characteristics of the stone. Conversely, a yellow dandelion flower growing between the marble's cracks signified the expression of unique life being allowed to emerge. As a methodological intervention, the experience of spending time in the field showed that dissonance can be utilised to reveal the way atmospheres attune us in certain ways. As I outlined in Chapters 2 and 5, personal experiences with obstructive staff and thieving sea birds disrupted my attunement to the OOH, enacting a kind of forced *withdrawal* from the atmosphere I was immersed in at the time. These encounters showed the workings of atmospheric attunement by ejecting me, at least temporarily, from its sphere. Moreover, the importance of dissonance was found to relate to the manner in which we feel connected to places and how we inhabit them in meaningful ways. In this regard, I have attempted to trouble the relationship between resonance and atmosphere, which has particular implications for the staging of place as it suggests that the smoothing of experience may not be as experientially *valuable* as the disruptions that dissonance provokes. Going forward, this poses the question of whether or not such ruptures in atmosphere are in fact a desirable goal for developers and designers, or whether the cohesion of urban experience takes precedence over meaningful engagement. The balance between resonance and dissonance requires further consideration and should be of interest to both designers and social scientists. The point at which meaningful experience is, in fact, dissonant, and the zone where resonant experience becomes coercively smoothed, has repercussions for the way atmospheres are formed and ruptured, and to what ends.

Ambiguity appears to be a necessary, if not intentional, part of the OOH's success, allowing different meanings and functions to overlap and co-exist. What this book underlines is that the way these tensions are smoothed and navigated, for example, in the OOH's status as a 'social monument', should be understood in terms of a staging of the atmospheric, helping us understand the implications of specific assemblages of material and immaterial elements as a particular distribution of the sensible. The forms of antagonism this book has identified are tacitly manifested by the OOH's design and its artistic contributors that drive visitors towards unity in the face of conflict. And in this sense, instead of sustaining heterogeneity, 'the house' constitutes a form of engagement which 'domesticates' (see Mouffe 1999, 754) through atmospheres of potential and futurity, staging a space for particular modes of participation to play out.

If the social monument is an outlier, a new take on cutting-edge architecture that establishes the social through meaningful experience, then the question becomes: how can it continue to constitute new ways of thinking the unthinkable? Once the 'social monument' has established its conditions, even as its own inception was once unthinkable, how will it be able to re-work its own sensible in the future? In other words, how can architecture sustain the conditions for dissensus to take place?

Such questions relate to urban development's relationship with more radical forms of democracy and whether works like the OOH can ever 'recompose political spaces, or if they must be content to parody them' (Rancière 2006, 92). The social monument presents us with a paradox that promises a consolidation of disparate terms. At the time of writing, that the social monument might turn out to be *a monument to the social*, does nothing to diminish its perceived qualities and the affection with which it is received by locals and visitors alike.

Bibliography

Adey, P. (2008) 'Airports, mobility and the calculative architecture of affective control', *Geoforum*. [Online], 39 (1), pp. 438–451.

Adey, P. (2015) 'Air's Affinities: Geopolitics, Chemical Affect and the Force of the Elemental', *Dialogues in Human Geography*, 5(1), pp. 54–75.

Ahmed, S. (2014) 'Not in the Mood', *New Formations*, 82, pp. 13–28.

Alaily-Mattar, N., Ponzini, D., and Thierstein, A. (2020) *About Star Architecture: Reflecting on Cities in Europe*. Nadia, Alaily-Mattar et al. (eds.). Cham: Springer International Publishing.

Albertsen, N. (2016) 'Atmosphere: Power, Critique, Politics. A Conceptual Analysis', Ambiances, tomorrow. Proceedings of the 3rd International Congress on Ambiances. Volos, Greece, September, pp. 573–578.

Albertsen, N. (2019) 'Urban Atmospheres', *Ambiances*, 24(3), pp. 353–374.

Al-Kodmany, K. (2020) *Tall Buildings and the City: Improving the Understanding of Placemaking, Imageability, and Tourism*. Singapore: Springer.

Allen, J. (2006) 'Ambient Power: Berlin's Potsdamer Platz and the Seductive Logic of Public Spaces', *Urban Studies*, 43(2), pp. 441–455.

Almaas, I.H. (2008) The sky's the limit: An introduction to the Oslo Opera. Available at: https://architecturenorway.no/stories/other-stories/iha-on-opera-2008/ Accessed (04 October 2020).

Alvesson, M., and Sköldberg, K. (2009) *Reflexive Methodology: New Vistas for Qualitative Research*. 2nd edition. London: SAGE.

Amin, A. (2008) Collective culture and urban public space. *City (London, England)*. [Online] 12 (1), 5–24.

Amin, A., and Thrift, N.J. (2017) *Seeing like a City*. Cambridge and Malden: Polity.

Anderson, B. (2019) 'Cultural geography II: The Force of Representations', *Progress in Human Geography*, 43(6), pp. 1120–1132.

Anderson, B., and Ash, J. (2015) 'Atmospheric Methods', in Vannini, P. (ed.), *Non-Representational Methodologies*. New York: Routledge, pp. 34–51.

Anderson, B., and Wylie, J. (2009) 'On Geography and Materiality', *Environment and Planning A*, 41(2), pp. 318–335.

Arnstein, S.R. (2019) 'A Ladder of Citizen Participation', *Journal of the American Planning Association*, 85(1), pp. 24–34.

Ash, J. (2015) Technology and affect: Towards a theory of inorganically organised objects. *Emotion, space and society*. [Online] 1484–1490.

Ash, J. & Simpson, P. (2016) 'Geography and post-phenomenology', *Progress in Human Geography*, 40(1), pp. 48–66.

Askheim, O.P. et al. (2017) 'User Participation in the Norwegian Welfare Context: An Analysis of Policy Discourses', *Journal of Social Policy*, 46(3), pp. 583–601.

Aspen, J. (2013) 'Oslo: The Triumph of Zombie Urbanism', in: Rodolphe el-Khoury and Edward Robbins (eds.), *Shaping the City: Studies in History, Theory and Urban Design*, London: Routledge.

Augestad, P. (2020) 'Nature in the history of physical education and in the lives of Norwegians today', in Tin, M. et al (eds.) *The Nordic Model and Physical Culture*. London: Routledge, pp. 36–50.

Baltzer, N. (2004) "The Cloudy Translucence Like that of Jade …': Atmospheric Architectural Photography," in Baltzer, N., and & Forster, K.W. (eds.) *Metamorph 9*. International Architecture Exhibition, exh. cat. La Biennale di Venezia.

Banham, R. (1960) *Theory and Design in the First Machine Age*. London: The Architectural Press.

Barnaby, A. (2017) *Light Touches: Cultural Practices of Illumination, 1780–1900*. New York: Routledge.

Barns, S. (2020) *Platform Urbanism Negotiating Platform Ecosystems in Connected Cities*. 1st edition. Singapore: Springer Singapore.

Barthes, R. (1982) *Camera Lucida: Reflections on Photography*. London: Vintage.

Batchen, G.(1997) *Burning with desire: the conception of photography*. Cambridge, Mass: The MIT Press.

Beech, D. (2008) 'Include me out!' *Art Monthly*, Available online: http://www.artmonthly.co.uk/magazine/site/article/include-me-out-by-dave-beech-april-2008 (Accessed 20/09/2021).

Beiguelman, G. (2020) 'Mobile Art: From the WAP Promises to the App Bubbles', in Hjorth, L., de Souza e Silva, A., and Lanson, K. (eds.) *The Routledge Companion to Mobile Media Art*. London: Routledge, pp. 34–45.

Bell, J.A. (2009) '"For Scientific Purposes a Stand Camera is Essential": Salvaging photographic histories in Papua', in Morton, C.A., and Edwards, E. (eds.) *Photography, Anthropology, and History: Expanding the Frame*. Farnham: Ashgate.

Benjamin, W. (1979) *Reflections: Essays, Aphorisms, Autobiographical Writings*. New York: Harvest/HBJ Books.

Benjamin, W. (1999) *The Arcades project*. Cambridge: Belknap Press.

Benjamin, W. (2015) *Illuminations*. London: Vintage.

Bennet, J. (2010) *Vibrant Matter: A Political Ecology of Things*. North Carolina: Duke University Press.

Berger, J. (2005) *Berger on Drawing*. Occasional Press.

Berglund, N. (2020) Opera House draws the most tourists. Available online: https://www.newsinenglish.no/2011/07/18/opera-house-draws-the-most-tourists/ (Accessed 10/10/2020).

Bergman, C.E. (2010) 'A Performativity of Nordic Space: The Tension between Ritual and Sincerity Re-Embodied through Each Performance of Sweden's Allsång på Skansen', *Ethnologia Europaea*, 40(2), pp. 77–89.

Bille, M. (2015) 'Hazy worlds: Atmospheric Ontologies in Denmark', *Anthropological Theory*, 15(3), pp. 257–274.

Bille, M., Bjerregaard, P., and Sørensen, T.F. (2015) 'Staging atmospheres: Materiality, Culture, and the Texture of the In-Between', *Emotion, Space and Society*, 15, pp. 31–38.

Bille, M., and Schwabe, S. (2023) *The atmospheric city*. Abingdon, England: Routledge.

Bille, M., and Sørensen, T.F. (2016) 'Into the Fog of Architecture', in Bille, M., and Sørensen, T.F. (eds.) *Elements of Architecture: Assembling Archaeology, Atmosphere and the Performance of Building Spaces*. London: Routledge, pp. 1–29.

Bishop, C. (2004) 'Antagonism and Relational Aesthetics', October, 110, pp. 51–79.

Bishop, C. (2005) *Installation Art: A Critical History*. London: Tate.

Bishop, C. (2006) *Participation*. Cambridge: The MIT Press.

Bishop, C. (2011) *Participation and Spectacle: where are we now?* Available online: http://dieklaumichshow.doragarcia.org/pdfs/Bishop.pdf (Accessed: 10 October 2023).

Bishop, C. (2012) *Artificial Hells: Participatory Art and the Politics of Spectatorship*. London: Verso.

Bishop, C. (2013) *Radical museology, or, What's "contemporary" in museums of contemporary art?* London: Koenig Books.

Bjerkeset, S., and Aspen, J. (2017) 'Private-Public Space in a Nordic Context: The Tjuvholmen Waterfront Development in Oslo', *Journal of Urban Design*, 22(1), pp. 116–132.

Bjørvika (2021) Facts and History about Bjørvika's development. Available at: https://www.bjorvikautvikling.no/portfolio-item/facts-and-history/ (Accessed: 26 April 2021).

Böhme, G. (1993) 'Atmosphere as the Fundamental Concept of a New Aesthetics', *Thesis Eleven*, 36(1), pp. 113–126.

Böhme, G. (2003) 'Contribution to the Critique of the Aesthetic Economy', *Thesis Eleven*, 73(1), pp. 71–82.

Böhme, G. (2013) 'The art of the stage set as a paradigm for an aesthetics of atmospheres', *Ambiances*. DOI: 10.4000/ambiances.315

Böhme, G. (2014) 'Urban Atmospheres: Charting New Directions for Architecture and Urban Planning', in Borch, C. (ed.) *Architectural Atmospheres: On the Experience and Politics of Architecture*. Basel: Birkhäuser, pp. 42–59.

Böhme, G. (2016). *The Aesthetics of Atmospheres*. In J.-P. Thibaud (ed.). London: Routledge.

Böhme, G. (2017). *Atmospheric Architectures: The Aesthetics of Felt Spaces*. London: Bloomsbury.

Bolt, B. (2000) 'Shedding Light For The Matter', *Hypatia*, 15(2), pp. 202–216.

Borch, C. (2008) Foam architecture: managing co-isolated associations. *Economy and society*. [Online] 37 (4), pp. 548–571.

Borch, C. (2010) 'Organizational Atmospheres: Foam, Affect and Architecture', *Organization*, 17(2), pp. 223–241.

Borch, C. (ed.) (2014) *Architectural Atmospheres: On the Experience and Politics of Architecture*. Basel: De Gruyter.

Boswijk, A., Thijssen T., and Peelen, E. (2008) *The Experience Economy: A New Perspective*. Amsterdam: Pearson Prentice Hall.

Bourriaud, N. (2002) *Relational aesthetics*. Dijon: Les Presse du Reel.

Boy, J., and Uitermark, J. (2016) 'How to Study the City on Instagram', *PloS ONE*, 11(6), e0158161.

Bremner, G.A. (2018) '"In Bright Tints… Nature's own formation": The Uses and Meaning of Marble in Victorian Building Culture', in Napoli, J.N., and Tronzo, W. (eds.) *Radical Marble: Architectural Innovation from Antiquity to the Present*. London: Routledge, pp. 72–91.

Bromgard, G. et al. (2014) 'Janteloven and the Expression of Pride in Norway and the United States', *The Journal of social psychology*. [Online] 154 (5), 375–378.

Brünner, M. (2015) *Constructing Atmospheres: Test Sites for an Aesthetics of Joy*. Baunach: Spurbuchverlag.

Bürger, P. (2014) *Theory of the Avant-Garde and Critical Literary Science*. Manchester: Manchester University Press.

Campbell, H. (2021) 'Looking at Photographs: Thinking about Architecture', in Troiani, I., and Ewing, S. (eds.) *Visual Research Methods in Architecture*. Bristol: Intellect Books Ltd, pp. 127–140.

Christensen, T. (2015) 'A new opera house: combining entrepreneurship, garbage can features and windows of opportunity', *International Review of Administrative Sciences*, 81(4), pp. 734–751.

Citton, Y. (2009) 'Political Agency and the Ambivalence of the Sensible', in Rockhill, G., and Watts, P. (eds.) *Jacques Rancière: History, Politics, Aesthetics*. Durham: Duke University Press, pp. 120–139.

Colomina, B. (1996) *Privacy and Publicity: Modern Architecture as Mass Media*. Cambridge: The MIT Press.

Cooke, R. (2015) *Olafur Eliasson: 'I am not special'*. Available at: https://www.theguardian.com/artanddesign/2015/jun/21/olafur-eliasson-i-am-not-special-interview-tree-of-codes-ballet-manchester/ (Accessed: 11 November 2021).

Le Corbusier (1987) *The Decorative Art of Today*. London: The Architectural Press.

Corner, J., and Hirsch, A.B. (2014) *Landscape Imagination: Collected Essays of James Corner 1990–2010*. New York: Princeton Architectural Press.

Cornwall, A. (2002) 'Making Spaces, Changing Places: Situating Participation in Development', *Working Paper Series 170*. Brighton: Institute of Development Studies.

Cornwall, A. (2004) 'Spaces for Transformation? Reflections on Issues of Power and Difference in Participation in Development', in Hickey, S., and Mohan, G. (eds.) *Participation: From Tyranny to Transformation? Exploring New Approaches to Participation in Development*. London: Zed Books, pp. 75–91.

Cruz, E.G. (2020) 'Mobile Street Photography: Continued, Collective, and Contested Decisive Moments', in Hjorth, L., de Souza Silva, A., and Lanson, K. (eds.) *The Routledge Companion to Mobile Media Art*. New York: Routledge, pp. 324–334.

Dagsavisen (1998) 'Klasseseire for Oslo Øst'. *Dagsavisen 23. april*, p. 9.

Dalsgaard, P. & Kortbek, K. (2009) 'Staging Urban Atmosphere In Interaction Design'. *Nordes: Nordic Design research*. Available at: https://archive.nordes.org/index.php/n13/article/view/51/42. Accessed: 23/09/2024

Debord, G. (2006) 'Towards a Situationist International//1957' in Bishop, C. (ed.) *Participation*. Cambridge: The MIT Press, pp. 96–101.

Debord, G. (2016) *Society of the Spectacle*. Detroit: Black & Red.

Degen, M., Melhuish, C., and Rose, G. (2017) 'Producing Place Atmospheres Digitally: Architecture, Digital Visualisation Practices and the Experience Economy', *Journal of Consumer Culture*, 17(1), pp. 3–24.

Degen, M., and Rose, G. (2022) *The New Urban Aesthetic: Digital Experiences of Urban Change*. London: Bloomsbury.

Denning, A. (2015) *Skiing into Modernity: A Cultural and Environmental History*. Berkeley: University of California Press.

Dennis, K., and Urry, J. (2009) *After the Car*. Oxford: Polity Press.

Dezeuze, A. (2010) 'An Introduction to the "Do-It-Yourself" Artwork', in Dezeuze, A. (ed.) *The "Do-It-Yourself" Artwork: Participation from Fluxus to New Media*. Manchester: Manchester University Press, pp. 1–21.

Dicks, B. et al. (2006) Multimodal ethnography. *Qualitative research: QR*. [Online] 6 (1), pp. 77–96.

Dorrian, M. (2014) 'Museum Atmospheres: Notes on Aura, Distance and Affect', *Journal of Architecture*, 19(2), pp. 187–201.

Dykers, C. (2012) 'SNOHETTA: Architecture of Humanity', *Queen's Quarterly*, 119(3), pp. 378–400.

Ebbensgaard, C.L. (2017) '"I Like the Sound of Falling Water, It's Calming": Engineering Sensory Experiences through Landscape Architecture', *Cultural Geographies*, 24(3), pp. 441–455.

Ebbensgaard, C.L. (2019) Making sense of diodes and sodium: Vision, visuality and the everyday experience of infrastructural change. *Geoforum*. [Online], 103 pp. 95–104.

Edensor, T. (2015a) 'Light Art, Perception, and Sensation', *The Senses & Society*, 10(2), pp. 138–157.

Edensor, T. (2015b) 'Light Design and Atmosphere', *Visual Communication*, 14(3), pp. 331–350.
Edensor, T. (2017) *From light to dark: daylight, illumination, and gloom*. 1st ed. Minneapolis: University of Minnesota Press.
Edensor, T., & Hughes, R. (2019) 'Moving through a Dappled World: The Aesthetics of Shade and Shadow in Place', *Social & Cultural Geography*, 22 (9), pp. 1307–1325.
Eliasson, O. (2001) *The Mediated Motion*. Bregenz: Kunsthaus Bregenz.
Eliasson, O. 2014) 'Atmospheres, Art, Architecture: A Conversation between Gernot Böhme, Christian Borch, Olafur Eliasson & Juhani Pallasmaa', in *Architectural Atmospheres*. [Online]. Berlin, Boston: DE GRUYTER. pp. 90–107.
Eliasson, O. (2016) *Olafur Eliasson: Nothingness is not Nothing at All*. Shanghai: Long Museum.
Eliasson, O. (2021c) *Studio Olafur Eliasson: The Kitchen, 2016*. Available at: https://olafureliasson.net/archive/publication/MDA118052/studio-olafur-eliasson-the-kitchen#slideshow (Accessed: 07 October 2021).
Eliasson, O. (2021d) *The other wall 2004–2008*. Available at: https://olafureliasson.net/archive/artwork/WEK100814/the-other-wall (Accessed: 10/10/2021).
Eliasson, O. (2021e) *Music wall, 2006*. Available at: https://olafureliasson.net/archive/artwork/WEK100611/music-wall#slideshow/ (Accessed: 10 October 2021).
Eliasson, O. & Irwin, R. (2008) 'Take Your Time: a Conversation'. In Grynsztejn, M. (2007) *Take your time [vol. 1]: Olafur Eliasson*. London: Thames & Hudson.
Eliasson, O., and Jensen, M.K. (2000) 'What we have in common is that we are different', in *Your Position Surrounded and Your Surroundings Positioned*, Katrina Brown and Olafur Eliasson (eds.), Dundee, vol. 1999, pp. 14–19.
Eliasson, O. cited in Kroksnes, A.(2011) 'Meeting the Wall', in Mortensen, J. *Visual Art in the Oslo Opera House*. Oslo: Press Publishing. pp. 120–128.
Engels-Schwarzpaul, T. (2017) 'Approaching Atmospheres: Translator's Introduction', in Böhme, G. *Atmospheric Architectures: The Aesthetics of Felt Spaces*. London: Bloomsbury, pp. 1–11.
ETC (2021) *ETC Mosaic creates a stunning show for Oslo Central Station*. Available at: https://www.etcconnect.com/About/News/ETC-Mosaic-creates-a-stunning-show-for-Oslo-Central-Station.aspx. (Accessed: 03/03/2021).
Falleth, E., and Saglie, I.-L. (2011) 'Democracy or Efficiency: Contradictory National Guidelines in Urban Planning in Norway', *Urban Research & Practice*, 4(1), pp. 58–71.
Filler, M. (2013) 'The Risk of Being too Nice', *The New York Review of Books*, 60 (2), p. 15.
Fiskum, I. (2003) *En by i endring. Bjørvika - porten til det nye Oslo*. In Fiskum, I. (ed.). Ernst & Young AS.
Fjordenhus (2020) *Fjordenhus, den første bygning tegnet af kunstneren Olafur Eliasson og hans atelier*. Available at: https://www.fjordenhus.dk/dk/fjordenhus/ (Accessed: 5 April 2020).
Fløgstad, K. (2010) 'The Unobtrusively Monumental', in Anglès, M. (2010) *In favour of public space: ten years of the European Prize for Urban Public Space*. Barcelona: Centre de Cultura Contemporània de Barcelona. pp. 60–65.
Florida, R. (2002) *The Rise of the Creative Class… and How It's Transforming Work, Leisure, Community and Everyday life*. New York: Basic Books.
Forty, A. (2012) *Concrete and Culture: A Material History*. London: Reaktion Books Ltd.
Foster, H. (1996) *The Return of the Real. The Avant-Garde at the End of the Century*. Cambridge, Massachusetts: The MIT Press.
Foster, H. (2003) *'Master Builder', in Design and Crime (And Other Diatribes)*. London & New York: Verso. pp. 27–42.
Foster, H. (2006) 'Chat Rooms//2004', in Bishop, C. (ed.) *Participation*. Cambridge: The MIT Press, pp. 190–195.

Foster, H. (2013) *The Art-Architecture Complex*. London: Verso Books.
Franck, K., and Stevens, Q. (2007) *Loose Space: Possibility and Diversity in Urban Life*. London: Routledge.
Franks, B. (2000). 'New Right/New Left: An Alternative Experiment in Freedom', in Hughes, J., and Sadler, S. (eds.) *Non-Plan: Essays on Freedom, Participation and Change in Modern Architecture and Urbanism*. London: Routledge, pp. 32–43.
Freudendal-Pedersen, M. (2007) Mobility, motility and freedom: the structural story as analytical tool for understanding the interconnection. *Schweizerische Zeitschrift für Soziologie*. 33 (1), 27–44.
Frichot, H. (2008) Olafur Eliasson and the Circulation of Affects and Percepts: In Conversation. *Architectural design*. 78 (3), 30–35.
Friluftshuset (2021) Available at: https://friluftshuset.dntoslo.no/ (Accessed: 21 May 2022).
Funcke, Bettina (2005) 'Against Gravity'. *Artforum International*, 11(5), pp. 27–29.
Gandy, M. (2017) 'Urban Atmospheres', *Cultural Geographies*, 24(3), pp. 353–374.
Gandy, M. (2019) 'Enlarging the urban orchestra: re-thinking current approaches to landscape architecture', in Braae, E. & Steiner, H. *Routledge Research Companion to Landscape Architecture*. 1st edition. Milton: Routledge. pp. 184–193.
Garnert, J. (2011) 'On the cultural history of Nordic Light and Lighting'. In Sørensen, N. et al. *Nordic light : interpretations in architecture*. 1. oplag. Stenløse: Dansk Center for Lys.
Gehl, J. (1987) *Life Between Buildings: Using Public Space*. New York: Van Nostrand Reinhold.
Gehl Architects (2007) *Byrom og Gaterom. temahefte Bjørvika*. Available online: https://www.bjorvikautvikling.no/wp-content/uploads/2018/03/Temahefte2.pdf. (Accessed: 23/08/2020).
Gehl, J., and Svarre, B. (2013) *How to Study Public Life*. Washington: Island Press.
Gherardi, F. et al. (2019) The deterioration of Apuan white marble in contemporary architectural context. *Journal of cultural heritage*.
Giulianotti, R. (2020) 'Afterword: The Nordic Model and Physical Cultures in the Global Context', in Tin, M.B. et al (eds.) *The Nordic Model and Physical Culture*. London: Routledge, pp. 235–245.
Given, L. (2012) 'Photographs in Qualitative Research', in Anon Photonovella and Photovoice (eds.). *The SAGE Encyclopedia of Qualitative Research Methods*.
Gonzalez, S. (2011) 'Bilbao and Barcelona "in Motion": How Urban Regeneration "Models" Travel and Mutate in the Global Flows of Policy Tourism', *Urban Studies*, 48(7), pp. 1397–1418.
Goodwin, M. (2016) *Architecture's Discursive Space: Photography*. PhD thesis. Aalto University.
Gormley, A. (2008). *Interview with Hans Ulrich Obrist*. Available at: https://www.antonygormley.com/resources/texts/in-conversation-with-hans-ulrich-obrist-2015 (Accessed: 12/09/2020).
Green, K., Sigurjónsson, T., and Skille, E.Å. (eds.) (2019) *Sport in Scandinavia and the Nordic Countries*. London: Routledge.
Griffero, T. (2023) 'Introduction: Spatial, authoritative, Felt-Bodily Resonant, Unintentional: Hermann Schmitz's atmospheres', in Griffero, T. (ed.) *Herman Schmitz: Atmospheres*. Mimesis International, pp. 9–26.
Gurholdt, K.P., and Haukeland, P.I. (2020) 'Scandinavian friluftsliv (outdoor life) and the Nordic model: passions and paradoxes', in Tin, M.B. *The Nordic Model and Physical Culture*. London & New York: Routledge.
Hammer, E. (2011) 'Beyond Iconic', in Mortensen, J. (ed.) *Visual Art in the Oslo Opera House*. Oslo: Press Publishing, pp. 88–100.
Hankins, T. L. & Silverman, R. J. (1999) *Instruments and the imagination*. Princeton, N.J: Princeton University Press.
Harman, G. (2020) *Art and objects*. Cambridge: Polity.

Haugdal, E. (2009) Et sosialt monument. Kunst og kultur. [Online] 92 (4), 216–229.
Hawkes, D. (2020) *The Environmental Imagination: Technics and Poetics of the Architectural Environment*. 2nd edition. London: Routledge.
Hawkins, H. (2013) *For Creative Geographies: Geography, Visual Arts and the Making of Worlds*. London: Routledge.
Heidegger, M. (1971) 'Building Dwelling Thinking', in *Poetry, Language, Thought*. Translated by A. Hofstadter. New York: Harper and Row.
Herzogenrath, B. (2009). 'The Weather of Music: Sounding Nature in the Twentieth and Twenty-first Centuries'. In Herzogenrath, B. (ed) *Deleuze/Guattari & ecology*. Basingstoke: Palgrave Macmillan.
Hofseth, M. (2008) 'The New Opera House in Oslo – a Boost for Urban Development?' *Urban Research & Practice*, 1(1), pp. 101–103.
Hörl, E. & Burton, J. E. (2017) *General Ecology: The New Ecological Paradigm*. 1st edition. [Online]. London: Bloomsbury Publishing Plc.
Howes, D. (2010) Response to Sarah Pink. *Social anthropology*. [Online] 18 (3), 333–336.
Hughes, J., and Sadler, S. (eds.) (2000) *Non-Plan: Essays on Freedom, Participation and Change in Modern Architecture and Urbanism*. London: Routledge.
Hüppauf, B., and Wulf, C. (eds.) (2009) *Dynamics and Performativity of Imagination: The Image Between the Visible and the Invisible*. New York: Routledge.
Ibler, M. (2014) *A New Golden Age: Nordic Architecture & Design*. Aarhus: Archipress M.
Ingold, T. (2000) *The Perception of the Environment: Essays on Livelihood, Dwelling and Skill*. 1st edition. [Online]. London: Routledge.
Ingold, T. (2007) Earth, sky, wind, and weather. *The Journal of the Royal Anthropological Institute*. 13 (S1), pp. S19–S38.
Ingold, T. (2008) Bindings against Boundaries: Entanglements of Life in an Open World. *Environment and planning A*. 40 (8), pp. 1796–1810.
Ingold, T. (2011) 'Landscape or weather-world?', in *Being Alive*. Routledge. pp. 139–148.
Ingold, T. (2016) 'Lighting up the atmosphere' in Bille, M. & Sørensen, T.F. (eds.) *Elements of Architecture: Assembling Archaeology, Atmosphere and the Performance of Building Spaces*. London: Routledge, pp. 1–29.
Ingold, T. (2017) Surface Visions. *Theory, culture & society*. [Online] 34 (7–8), pp. 99–108.
Ingram, S. (2019) *Why This Artist is Bringing the Outdoors Indoors, and Unreality into Real Life*. Available at: https://www.nationalgeographic.co.uk/environment-and-conservation201907why-artist-bringing-outdoors-indoors-and-unreality-real-life/(Accessed: 17 May 2021).
Jacobs, J. (1961) *The Death and Life of Great American Cities*. New York: Random House.
Janson, A., and Tigges, F. (2014) *Fundamental Concepts of Architecture: The Vocabulary of Spatial Situations*. Basel: De Gruyter.
Jasper, S. (2020) 'Acoustic Ecologies: Architecture, Nature, and Modernist Experimentation in West Berlin', *Annals of the American Association of Geographers*, 110(4), pp. 1114–1133.
Jencks, C. (2006) 'The Iconic Building is Here to Stay', *City*, 10(1), pp. 3–20.
Jenkins, P. (2009) 'Concepts of Social Participation in Architecture', in Jenkins, P. & Forsyth, L. (eds.) *Architecture, Participation, and Society*. London: Routledge, pp. 9–22.
Jones, O. (2009) 'Dwelling', in Kitchin, R., and Thrift, N. (eds.) *International Encyclopedia of Human Geography*. Oxford: Elsevier Science & Technology, pp. 266–272.
Jørgensen, F.A. (2014) 'The Networked North: Thinking about the Past, Present, and Future of Environmental Histories of the North', in Jørgensen, D., and Sörlin,

S. (eds.) *Northscapes: History, Technology, and the Making of Northern Environments*. Vancouver: UBC Press, pp. 268–279.

Jørgensen, L., and Borch, C. (2018) 'Arkitektur og politik: Atmosfæriske refleksioner ud fra Böhme og Sloterdijk', *Politik*, 21(2), pp. 7–21.

Jormakka, K. (2004) 'The most architectural thing', in Mical, T. (ed.) *Surrealism and Architecture*. United Kingdom: Routledge. pp. 304–331.

Jurgenson, N. (2020) *The Social Photo: On Photography and Social Media*. New York: Verso Books.

Kaika, M. (2011) 'Autistic Architecture: The Fall of the Icon and the Rise of the Serial Object of Architecture', *Environment and Planning D*: Society & Space, 29(6), pp. 968–992.

Kaika, M., and Thielen, K. (2006) 'Form Follows Power: A Genealogy of Urban Shrines', *City*, 10(1), pp. 59–69.

Kelling, G., and Wilson, J. (1982) 'Broken Windows', *the Atlantic*, Available at: https://web.archive.org/web/20090418141450/; http://www.theatlantic.com//doc//198203//broken-windows (Accessed: 18 March 2021).

Kiib, H. (2007) *Harbourscape*. Aalborg: Aalborg University Press.

Klingmann, A. (2007) *Brandscapes: Architecture in the Experience Economy*. Cambridge: The MIT Press.

Koolhaas, Rem. "Junkspace." October, vol. 100, 2002, pp. 175–190. JSTOR, http://www.jstor.org/stable/779098. Accessed 8 Sept. 2024.

Kortbek, H.B. et al. (2017) 'The Participatory Agenda: A Post-Critical, Anticipatory Intervention', *Nordisk Kulturpolitisk Tidskrift*, 19(1), 19–35.

Kotler, P. (1973) 'Atmospherics as a Marketing Tool', *Journal of Retailing*, 49, pp. 48–64.

Kozinets, R., Gretzel, U., and Dinhopl, A. (2017) 'Self in Art/Self As Art: Museum Selfies As Identity Work', *Frontiers in Psycholology*, 8, 731.

Kracauer, S. (1995) *The Mass Ornament: Weimar Essays*. Translated by T.Y. Levin. Cambridge: Harvard University Press.

Kroksnes, A. (2011) 'Meeting the Wall', in Mortensen, J. (ed.) *Visual Art in the Oslo Opera House*. Oslo: Press Publishing.

Larsen, J. (2004) (Dis)Connecting Tourism and Photography: Corporeal Travel and Imaginative Travel. *Journeys (New York, N. Y.)*. [Online] 5 (2), pp. 19–42.

Larsen, B. (2006). In Bishop, C. *Participation*. Cambridge, Mass: The MIT Press.

Larsen, J. (2021) *Urban Marathons: Rhythms, Places, Mobilities*. London: Routledge.

Latham, A., and McCormack, D.P. (2009) 'Thinking with Images in Non-Representational Cities: Vignettes from Berlin', *Area*, 41(3), pp. 252–262.

Lavender, A. (2012) 'Viewing and Acting (and Points in Between): The Trouble with Spectating after Rancière', *Contemporary Theatre Review*, 22(3), pp. 307–326.

Leach, N. (ed.) (1997) *Rethinking Architecture: A Reader in Cultural Theory*. New York: Routledge.

Leavy, P. (2015) *Method Meets Art: Arts-Based Research Practice*. 2nd edition. New York: The Guilford Press.

Lidwell, W., Holden, K., and Butler, J. (2010) *Universal Principles of Design*. Beverly: Rockport.

Loos, A. (1908) 'Ornament and Crime', in Conrads, U. (ed.) (1971) *Programs and Manifestoes on 20th-Century Architecture*. 16. print. Cambridge: The MIT Press, pp. 19–25.

Lorimer, H. (2012) 'Surfaces and Slopes', *Performance Research*, 17(2), pp. 83–86.

Löschke, S.A. (2016) *Materiality and Architecture*. London: Routledge.

Lundevall, T. (2008) 'Prosjektering og byggeprosess', in Opera Forlag Jon Otterbeck AS Bogstadveien 11. 0355 Oslo, pp. 58–81.

Lyons, F. (2019) *The Architecture of Nothingness: An Explanation of the Objective Basis of Beauty in Architecture and the Arts*. London: Routledge.

MacDowall, L., and Budge, K. (2021) *Art after Instagram: Art Spaces, Audiences, Aesthetics*. London: Routledge.
MacDowall, L. & Budge, K. (2022) *Art after Instagram: Art Spaces, Audiences, Aesthetics*. London: Routledge.
MacKeith, P. (2008) National Opera of Norway. Architectural record. 196 (8), 84.
MacKeith, P. (2012) 'The building art the social art: Reflections on a Nordic social architecture', in Kjeldsen, K. (ed.) *New Nordic: Architecture & Identity*. Humlebæk: Louisiana Museum of Modern Art, pp. 134–155.
Madsen, M. & Kristensen, P. (2006) 'Daylight as a Building Material', in *Lighting Design*. [Online]. Switzerland: Walter de Gruyter GmbH. pp. 92–101.
Malpas, J. (2021) *Rethinking dwelling: Heidegger, place, architecture*. [Online]. London; Bloomsbury Academic.
Mansfield, K. (2018) Architectural lighting design: A research review over 50 years. *Lighting research & technology (London, England: 2001)*. 50 (1), 80–97.
Marfelt, B. (2007) *Oslos hvide marmoropera er ved at blive tis-gul*. Available at: https://ing.dk/artikel/oslos-hvide-marmoropera-er-ved-blive-tis-gul-83579/ (Accessed: 20 November 2021).
Marklund, C. (2014) 'From Promise to Compromise: Nordic Openness in a World of Global Transparency'. In Götz, N. & Marklund, C. *The paradox of openness: transparency and participation in nordic cultures of consensus*. Leiden: Brill.
Marling, G., Jensen, O.B., and Kiib, H. (2009) 'The Experience City: Planning of Hybrid Cultural Projects', *European Planning Studies*, 17(6), pp. 863–885.
Marshall, R. (2001) *Waterfronts in Post-Industrial Cities*. New York: Spon Press.
Marshall, R. (2007) *Waterfronts in Post-Industrial Cities*. New York: Spon Press.
Massey, D. 2005. *For Space*. London: Sage.
McCarthy, J., and Wright, P. (2015) *Taking [A]part: The Politics and Aesthetics of Participation in Experience-Centered Design*. Cambridge: The MIT Press.
McCormack, M.P. (2014) *Refrains for Moving Bodies: Experience and Experiment in Affective Spaces*. Durham: Duke University Press.
McCormack, D.P. (2018) *Atmospheric Things: On the Allure of Elemental Envelopment*. Durham: Duke University Press.
McCullough, M. (2013) *Ambient Commons: Attention in the Age of Embodied Information*. Cambridge: The MIT Press.
Michels, C. (2015) Researching affective atmospheres. *Geographica Helvetica*. [Online] 70 (4), pp. 255–263.
Miles, M. (1997) *Art, Space and the City: Public Art and Urban Futures*. London: Routledge.
Miles, M. (2015) *Limits to Culture: Urban Regeneration vs. Dissident Art*. London: Pluto Press.
Miller-Lane, B. (2000) *National Romanticism and Modern Architecture in Germany and the Scandinavian Countries*. Cambridge: Cambridge University Press.
Moran, J. (2006) *Interdisciplinarity*. London. Routledge.
Morrison-Low, A.D. (2005) 'Pre-Photographic Imaging Devices', in Lenman, R., and Nicholson, A. (eds.) *The Oxford Companion to the Photograph*. Oxford: Oxford University Press.
Mortensen, J. (ed.) (2011) *Visual Art in the Oslo Opera House*. Oslo: Press Publishing.
Morton, T. (2010) *The Ecological Thought*. Cambridge: Harvard University Press.
Morton, T. (2018) *Being Ecological*. London: Pelican.
Mouffe, C. (1999) 'Deliberative Democracy or Agonistic Pluralism?' *Social research*. 66 (3), pp. 745–758.
Mouffe, C. (2005) 'Some Reflections on an Agonistic Approach to the Public', in Latour, B. et al. *Making things public: atmospheres of democracy*. Cambridge, Mass: MIT Press.

Nielsen, T. (2012) 'Democratic urban spaces in the nordic countries: informal co-existence and amenities for all', in Kjeldsen, K. (ed.) *New Nordic: Architecture & Identity*. Humlebæk: Louisiana Museum of Modern Art, pp. 166–186.
Nielsen, T. (2019) 'The Making of Democratic Urban Public Space in Denmark', in Aelbrecht, P., and Stevens, Q. *Public Space Design and Social Cohesion: An International Comparison*. 1st edition. Vol. 1. Milton: Routledge. pp. 37–57.
Norberg-Schultz, C. (1996) *Nightlands: Nordic Building*. Cambridge: The MIT Press.
Norberg-Schulz, C. (1984) *Genius Loci: Towards a Phenomenology of Architecture*. New York: Rizzoli.
Nørretranders, T., and Eliasson, O. (2015) *Lys! Om lys i livet og liv i lyset*. Klampenborg: Forlaget Tor.dk.
Norwayexperts (2021). *Blue, green and the city in between*. Available at: https://www.norwayexports.no/news/blue-green-and-the-city-in-between/ Accessed: (10 November 2021).
NRK (2014) *Advarer mot mini-Manhattan*. Available at: https://www.nrk.no/osloogviken/advarer-mot-mini-manhattan-1.11493591 (Accessed 08/08/2020).
Nytt operahus i Oslo /New Opera House in Oslo (2002*) Open international project competition for Architects: the panels assessments and conclusions*.
Olin, M. (2008) *A New House in the City*. Available at: https://www.architecturenorway.no/stories/other-stories/olin-on-opera-2008/(Accessed: 25 October 2021).
Opprann, K. (2008) *Vi Bygde Operaen*. Oslo: Font forlag.
Oslohavn (2021) *Fjord City*. Available at: https://www.oslohavn.no/en/menu/fjordbyen/ (Accessed: 16 September 2024).
Otter, C. (2008) *The Victorian eye: a political history of light and vision in Britain, 1800-1910*. 1st ed. Chicago: University of Chicago Press.
Otterbeck, J. (2008) *Oslo Opera House*. Oslo: Opera Forlag.
Pallasmaa, J. (2011) 'An Architecture of the Seven Senses', In Weinthal, L. (ed.), *Toward a New Interior*. New York: Princeton Architectural Press.
Pallasmaa, J. (2012) *The Eyes of the Skin: Architecture and the Senses*. 3rd edition. Chichester: Wiley.
Pallasmaa, J. (2014) 'Space, Place, and Atmosphere: Pheripheral Perception in Existential Experience', in Borch, C. (ed.) *Architectural Atmospheres: On the Experience and Politics of Architecture*. Basel: Birkhäuser, pp. 18–41.
Palmer, D. (2020) 'Mobile Photography and Artistic Activism in the "Instagram Museum"', in Hjorth, L., de Souza e Silva, A., and Lanson, K. (eds.) *The Routledge Companion to Mobile Media Art*. London: Routledge, pp. 313–323.
Parent, C., and Virilio, P. (1996) *The Function of the Oblique: The Architecture of Claude Parent and Paul Virilio 1963–1969*. In P. Johnston (ed.). London: AA Publications.
Payne-Frank, J., and Schwabe, S. (2022) Staging Openness through Atmosphere at the Oslo Opera House. *Architecture and culture*, 10 (1), 39–57.
Pérez-Gómez, A. (2017) 'Architecture and the City: The Space of Participation', *Studii de istoria şi teoria arhitecturii*, 5, pp. 13–18.
Philippopoulos-Mihalopoulos, A. (2013) 'Atmospheres of Law: Senses, Affects, Lawscapes', *Emotion, Space and Society*, 7, pp. 35–44.
Philippopoulos-Mihalopoulos, A. (2015) *Spatial Justice: Body, Lawscape, Atmosphere*. London: Routledge.
Philippopoulos-Mihalopoulos, A. (2016) 'Withdrawing from Atmosphere: An Ontology of Air Partitioning and Affective Engineering', *Environment and Planning D: Society & Space*, 34(1), pp. 150–167.
Pinder, D. (2011) 'Errant Paths: The Poetics and Politics of Walking', *Environment and Planning D: Society & Space*, 29(4), pp. 672–692.
Pine, B.J., and Gilmore, J.H. (2011) *The Experience Economy*. Boston: Harvard Business Review Press.

Pink, S. (2008) 'An Urban Tour: The Sensory Sociality of Ethnographic Place-Making', *Ethnography*, 9(2), pp. 175–196.
Pink, S. (2015) *Doing Sensory Ethnography*. 2nd edition. London: Sage.
Platt, L., Medway, D., and Steadman, C. (2021) 'Processional Walking: Theorising the "Place" of Movement in Notions of Dwelling', *Geographical Research*, 59(1), pp. 106–117.
Pløger, J. (2021) 'Politics, Planning, and Ruling: The Art of Taming Public Participation', *International Planning Studies*, 26(4), pp. 426–440.
Plummer, H. (1987) *Poetics of light*. Tokyo: A+U Publishing.
Plummer, H. (2012) *Nordic Light: Modern Scandinavian Architecture*. London: Thames & Hudson.
Ponzini, D. (2012) 'Competing Cities and Spectacularizing Urban Landscapes', in Anheier, H.K., and Isar, Y.R. *Cities, Cultural Policy and Governance*. Los Angeles: SAGE, pp. 99–110.
Ramboll (2020) *Oslo Opera House*. Available at: https://ramboll.com/projects/rno/operahuset%20i%20oslo (Accessed: 10 November 2021).
Rancière, J. (2002) 'The aesthetic revolution and its outcomes: emplotments of autonomy and heteronomy', *New Left review*. (14), pp. 133–151.
Rancière, J. (2006) 'Problems and Transformations in Critical Art'. In Bishop, C. (2006) *Participation*. Cambridge, Mass: MIT Press.
Rancière, J. (2009) *The Aesthetic Unconcious*. Cambridge: Polity.
Rancière, J. (2010) The Paradoxes of Political Art, in *Dissensus: On Politics and Aesthetics*. London: Continuum.
Rancière, J. (2011) *The Emancipated Spectator*. Translated by G. Elliott. London: Verso.
Rancière, J. (2013) *The politics of aesthetics: the distribution of the sensible*. Edited and translated by G. Rockhill. Bloomsbury Revelations edition. London: Bloomsbury Academic.
Rancière, J. (2019a) *Aisthesis: Scenes from the Aesthetic Regime of Art*. Translated by Z. Paul. London: Verso.
Rancière, J. (2019b) *The Future of the Image*. Translated by G. Elliott. London: Verso.
Rauh, A. (2018) *Concerning Astonishing Atmospheres: Aisthesis, Aura and Atmospheric Portfolio*. Mimesis International.
Rendell, J. (2002) 'Where the Thinking Stops, Time Crystallises …', in Miles, M., and Hall, T. (eds.) *Urban Futures: Critical Commentaries on Shaping the City*. London: Routledge, pp. 13–26.
Rendell, J. (2006) *Art and architecture: a place between*. London: I B Tauris & Co Ltd.
Retallack, J. (2004) *The Poethical Wager*. Berkeley: University of California Press.
Rexer, L. (2009) *The Edge of Vision: The Rise of Abstraction in Photography*. New York: *Aperture*.
Rexer, L. (2011) 'Knowing where to stand'. *Metropolis* (New York, N.Y.). 31 (4), 44–49.
Richardson, T., and Connelly, S. (2012) 'Reinventing public participation: planning in the age of consensus', in Blundell-Jones, P. et al. *Architecture and participation*. 1st edition. London: Routledge.
Riedel, F. (2020) *Music as Atmosphere: Collective Feelings and Affective Sounds*. Friedland, Riedel, and Juha Torvinen (eds.). Routledge
Risom, J. (2016) 'The Nordic Urban Design Model', in Simonyi, A., and Cagan, D.L. (eds.) *Nordic Ways*. Washington, DC: Center for Transatlantic Relations, pp. 214–222.
Roberston, T., and Simonsen, J. (2012) 'Participatory Design: An Introduction', in *Routledge International Handbook of Participatory Design*. London: Routledge, pp. 21–38.
Rockhill, G.(2009) *Jacques Rancière: history, politics, aesthetics*. Durham: Duke University Press.

Rockhill, G. (2016) *Interventions in Contemporary Thought: History, Politics, Aesthetics*. Edinburgh: Edinburgh University Press.

Rogoff, I. (2002) 'WE. Collectivities, Mutualities, Participations', in von Hantelmann, D., and Jongbloed, M. (eds.) *I promise It's Political: Performativät in der Kunst*. Bonn: Theater der Welt.

Rosa, H. (2020) *The Uncontrollability of the World*. 1st edition. Newark: Polity Press.

Rose, G. (2016) *Visual Methodologies: An introduction to Researching with Visual Materials*. London, Sage.

Rosenfield, K. (2021) *Astrup Fearnley Museet/Renzo Piano Building Workshop*. Available at: https://www.archdaily.com/282370/astrup-fearnley-museet-renzo-piano-building-workshop/ (Accessed: 05 April 2021).

Rudestam, K.E., and Newton, R.R. (2007) *Surviving your Dissertation: A Comprehensive Guide to Content and Process*. Los Angeles: SAGE.

Schinkel, W. & Noordegraaf-Eelens, L. (2011) *In Medias Res: Peter Sloterdijk's Spherological Poetics of Being*. 1st edition. Vol. 45175. [Online]. Amsterdam: Amsterdam University Press.

Schønsby (2011) 'Built- in a backward glance', in Mortensen, J. (ed.) *Visual Art in the Oslo Opera House*. Oslo: Press Publishing, pp. 136–145.

Schumacher, P. (2018) 'Freedom Via Soft Order: Architecture as a Foil for Social Self-Organisation', *Architectural Design*, 88(3), pp. 76–83.

Schwabe, S. (2021) 'Order and Atmospheric Memory: Cleaning Up the Past, Designing the Future', *City & Society*, 33(1), pp. 40–58.

Seippel, Ø., and Skille, E.Å. (2019) 'Sports Participation in Norway', in Green, K., Sigurjónssonn, T. & Skille, E.Å. (eds.) *Sport in Scandinavia and the Nordic Countries*. London: Routledge, pp 108–135.

Sennett, R. (2006) *'The Open City,'* available at: https://urbanage.lsecities.net/essays/the-open-city. Accessed: 10 June 2022.

Sennett, R. (2018) *Building and Dwelling: Ethics for the City*. London: Penguin.

Sennett, R., and Sendra, P. (2020) *Designing Disorder: Experiments and Disruptions in the City*. London: Verso.

Shanks, M., and Svabo, C. (2013) 'Mobile-Media Photography: New Modes of Engagement', in Larsen, J., and Sandbye, M. (eds.) *Digital Snaps: The New Face of Photography*. London: Routledge, pp. 227–246.

Sheller, M., Kesselring, S., and Jensen, O.B. (2019) 'Introduction', in Jensen, O.B., Kesselring, S., and Sheller, M. (eds.) *Mobilities and Complexities*. London: Routledge, pp. 1–4.

Sklair, L. (2010) 'Iconic Architecture and the Culture-ideology of Consumerism', *Theory, Culture & Society*, 27(5), pp. 135–159.

Skrede, J. (2016) 'What May Culture Contribute to Urban Sustainability? Critical Reflections on the Uses of Culture in Urban Development in Oslo and Beyond', *Journal of Urbanism*, 9(4), pp. 408–425.

SLA (2020) *Bjørvika masterplan and urban spaces*. Available at: https://www.sla.dk/cases/bjorvika-masterplan-and-urban-spaces/ (Accessed: 05 April 2021).

Sloterdijk, P. (2009) 'Airquakes', *Environment and Planning D: Society & Space*, 27(1), pp. 41–57.

Sloterdijk, P. (2014) *In the World Interior of Capital: For a Philosophical Theory of Globalization*. Translated by W. Hoban. Cambridge: Polity Press.

Sloterdijk, P. (2016). *Spheres Volume 3: Foams, Plural Spherology*. South Pasadena: Semiotext(e).

Smith, A., and von Krogh Strand, I. (2011) 'Oslo's New Opera House: Cultural Flagship, Regeneration Tool or Destination Icon?', *European Urban and Regional Studies*, 18(1), pp. 93–110.

Snøhetta (2021d) *Bibliotheca Alexandrina*. Available at: https://snohetta.com/project/5-bibliotheca-alexandrina(Accessed: 11 November 2021).

Solnit, R. (2014) *Wanderlust: A History of Walking*. London. Granta.

Sontag, S. (1979) *On Photography*. London. Penguin Books.

Sørensen, T.F. (2014). 'More than a Feeling: Towards an Archaeology of Atmosphere', *Emotion, Space and Society*, 15, pp. 64–73.

Sørensen, A. (2016). '"Participation": The New Cultural Policy and Communication Agenda', *Nordisk Kulturpolitisk Tidskrift*, 19(1), pp. 4–18.

Stanek, L. (2012) 'Architecture as Space, Again? Notes on the 'Spatial Turn''. *SpecialeZ*. 4, pp. 48–53.

Statsbygge (1994) *Nytt operagygg i Oslo: konsekvensutredning etter plan og bygningsloven. kpt*. VII-a.

Statsbygge (2005) *New Opera House: Presentation of the project New Opera House*.

Stenslund, A. (2015) 'A Whiff of Nothing: The Atmospheric Absence of Smell', *The Senses & Society*, 10(3), pp. 341–360.

Stevens, Q. (2007). 'Betwixt and Between: Building Thresholds, Liminality and Public Space', in Franck, K., and Stevens, Q. *Loose Space: Possibility and Diversity in Urban Life*. London: Routledge, pp. 73 – 92.

Stevens, Q., and Lossau, J. (2015). 'Framing Art and Its Uses in Public Space', in Lossau, J., and Stevens, Q. (eds.) *The Uses of Art in Public Space*. London: Routledge. pp. 1–18.

Stewart, K. (2007) *Ordinary affects*. Durham, NC: Duke University Press.

Steyerl, H. (2013) In defence of the poor image. *Konteksty*. 67 (3), 101–105.

Sumartojo, S., and Pink, S. (2018a) 'Moving Through the Lit World: The Emergent Experience of Urban Paths', *Space and Culture*, 21(4), pp. 358–374.

Sumartojo, S., and Pink, S. (2018b) *Atmospheres and the Experiential World: Theory and Methods*. London: Routledge.

Sumartojo, S., Pink, S., Edensor, T. (2019) 'Atmospheres in Urban Light', *Ambiances*, 5(5).

Tanke, J. J. (2010) Why Rancière Now? *The Journal of aesthetic education*. [Online] 44 (2), 1–17.

Tate ETC (2019) *Olafur Eliasson talks to Lily Cole*. Available at: https://www.tate.org.uk/tate-etc/issue-46-summer-2019/olafur-eliasson-talks-lily-cole (Accessed: 03 March 2021).

Thibaud, J.-P. (2015) 'The Backstage of Urban Ambiances: When Atmospheres Pervade Everyday Experience', *Emotion, Space and Society*, 15, pp. 39–46.

Thorsen, K.T., and Obrist, U. (2011) 'Long Duration', in Mortensen, J. (ed.) *Visual Art in the Oslo Opera House*. Oslo: Press Publishing, pp. 46–66.

Till, J. (2005) 'The Negotiation of Hope', in Jones, P.B., Petrescu, D., and Till, J. (eds.) *Architecture and Participation*. London: Routledge, pp. 23–41.

Tingsby, B. (2016) *Nordic Voices: The Global Voice of the Nordic Region*. Copenhagen: Nordic Council of Ministers.

Toffler, A. (1980) *The Third Wave*. New York: William Morrow.

Tostrup, E. (2009) Tracing competition rhetoric. *Nordic Journal of Architectural Research*.

Trägårdh, L. (2012) 'Nordic Modernity: Social Trust and Radical Individualism', in Kjeldsen, K. (ed.) *New Nordic: architecture & identity*. 1. oplag. Humlebæk: Louisiana Museum of Modern Art.

Tuan, Y.-F. (2014) *Space and Place: The Perspective of Experience*. Minneapolis: University of Minnesota Press.

Turner, F. (2013) *The democratic surround: multimedia & American liberalism from World War II to the psychedelic sixties*. Chicago: The University of Chicago Press.

Turner, C. (2018) 'Drawing, Adrift: Bengaluru – Mumbai – St Ives', *Performance Research*, 23(7), pp. 36–44.

Urry, J., and Larsen, J. (2011). *The Tourist Gaze 3.0*. Los Angeles: SAGE.

Vadalá, D. (2018) 'The Silent Seed of Modernity: How Marble Made Rationalism', in Napoli, J.N., and Tronzo, W. (eds.) *Radical Marble: Architectural Innovation from Antiquity to the Present*. London: Routledge, pp. 93–106.
Valtysson, B. (2016) 'Not Cool, but Cosy – Five Perspectives on Nordic City Cultural Policies', *Nordisk Kulturpolitisk Tidskrift*, 2, pp. 183–200.
Vannini, P. (2019) *Doing Public Ethnography: How to Create and Disseminate Ethnographic and Qualitative Research to Wide Audiences*. London: Routledge.
Vannini, P., and Vannini, A. (2020) 'Attuning to Wild Atmospheres: Reflections on Wildness as Feeling', *Emotion, Space and Society*, 36, 100711.
Vergunst, J.L. (2016) 'Taking a Trip and Taking Care in Everyday Life', in Ingold, T. & Vergunst, J.L. (eds.) *Ways of Walking: Ethnography and Practice on Foot*. London: Routledge, pp. 105–121.
Vibe, E. de (2007) 'Fjordbyens visjoner, aktører og prosesser', *Plan*, 39(1), pp. 32–39.
Vollsnes, A.O. et al. (2010) *Norges opera & balletthistorie*. Oslo: Opera Forlag.
Wainwright, O. (2012) *Renzo Piano's new Oslo art gallery is a strangely soulless place*. Available at: https://www.theguardian.com/artanddesign/architecture-designblog/2012/oct/11/renzo-piano-oslo-astrup-fearnley-art-museum. (Accessed: 17 February 2021).
Warren, S. (2012) 'Having an Eye for It: Aesthetics, Ethnography and the Senses', *Journal of Organizational Ethnography*, 1(1), pp. 107–118.
Ween, G., and Abram, S. (2012) 'The Norwegian Trekking Association: Trekking as Constituting the Nation', *Landscape Research*, 37(2), pp. 155–171.
Weidinger, J. (ed.) (2018). *Designing Atmospheres*. Berlin: Universitätsverlag der TU Berlin.
Weiss, S. (2020) 'Between Social Engagement and Neoliberalism', in Löschke, S.K. (ed.) *Non-Standard Architectural Productions: Between Aesthetic Experience and Social Action*. London: Routledge, pp. 235–252.
Wigley, M. (1996) *White Walls, Designer Dresses: The Fashioning of Modern Architecture*. Cambridge: The MIT Press.
Wigley, M. (2020) Chronic Whiteness. Available at: https://www.e-flux.com/architecture/sick-architecture/360099/chronic-whiteness/. (Accessed: 17 February 2021).
Wilk, R. (2005) 'Smoothing'. *Ethnologia Europaea*. 35 (1).
Worpole, K. (2000). *Here Comes the Sun: Architecture and Public Space in Twentieth-Century European Culture*. London: Reaktion Books.
Wylie, J. (2002) 'An Essay on Ascending Glastonbury Tor', *Geoforum*, 33(4), pp. 441–454.
Yarrow, T. (2019) *Architects: Portraits of a Practice*. Ithaca: Cornell University Press.
Žižek, S. (2013) 'The Lesson of Rancière', in Rancière, J. (ed.) *The Politics of Aesthetics: The Distribution of the Sensible*. Edited and Translated by G. Rockhill. London: Bloomsbury, pp. 65–75.
Zumthor, P. (2006) *Atmospheres: Architectural Environments, Surrounding Objects*. Basel: Birkhäuser Verlag AG.

Index

ÅF 119–120
aisthesis 5, 28, 152
Aker Brygge 47
Allen, J. 122, 153
Arnstein, S. 41, 57–58
art and architecture 13, 69, 121, 127
art as process 135–137
art autonomous 129–130, 146
atmosphere as method 6, 8, 14, 16, 18–23
attunement 4, 19–20, 28–29, 96

Benjamin, W. 20–22, 26
Bilbao effect 12, 47, 87, 103
Bille, M. 106
bipolar emotional response test (BERT) 32
Bishop, C. 126–127, 129–130, 135, 138, 140, 144, 147, 150
Bjørvika 39, 42–48, 54–55
Böhme, G. 8–9, 13–14, 19–31, 62–64, 69, 74–80, 108–113, 151–152
Bolt, B. 78–80
broken windows theory 122–123
bubbles 19, 26, 28, 123

Cage, J. 134–135

dark light 115–116
Degen, M. & Rose, G. 22, 29, 127, 141
Deichman library 39, 44
democratic architecture 12, 49
development process 57–59, 63
dissensus 129, 132, 145, 147, 155
dissonance 15, 32, 92, 94, 102–103, 153–155
dugnad 102
dwelling 24–25, 86, 92–93

ecological 77, 112, 128, 130–131, 135–137, 146
electric scooters 67, 99–100
Eliasson, O. 108–109, 117–119, 129–134, 136–137, 139–140, 144–147
emancipation of the spectator 4, 24, 133, 145
experience economy 13, 22, 26

Fjord city 1, 11, 37, 43–45, 62, 73
foam 14, 19, 26, 28–29, 107, 118, 123–124

Gehl architects 149
Gehl, Jan 10–11
glass 52, 87, 105, 113, 116, 122–123
Grønland 121, 124
guides 32, 54
gulls 83, 96, 103, 111, 113

Hammer, E. 84, 90, 95, 97–98, 101–103
Heidegger, M. 86
horizontality 85–87, 89–90, 151
hyper-lapse 33

immaterial architecture 11, 13, 23, 63, 73
Ingold, T. 25, 86, 93–94, 96, 110
inhabitation 14, 86–87, 92–94, 103
Instagram 7–8, 111, 123, 136, 140, 144

Janteloven 51, 60, 149

Le Corbusier 63, 67, 71, 80–81
light, liquid 118–120, 124
Light, Nordic 14, 49, 74, 106–107, 109, 114, 117, 119–120, 124
Lovaas & Wagle 73

marble 14, 42, 55–57, 62–81
mishaps 92–96, 102–103
mobility 86, 121, 149
Morton, T. 134–137
motile 99, 101, 103
multimodality 5, 7, 22, 29
Munch Museum 39, 44

Norberg-Schultz, C. 68, 77, 91, 115
nordic nature 65, 74, 81, 84
nordicness 7, 51
Norwegianness 52, 93, 101–103
Nolli, G. 51

openness 1, 18, 37, 44, 49, 54, 73, 86, 91, 100, 107, 109–111, 113, 117–118

Pallasmaa, J. 24, 80, 89, 93, 152
Parent, C. & Virilio, P. 88–89
Philippopoulos-Mihalopoulos, A. 8, 18–19, 21, 29, 113, 123–124
photography as method 8, 15, 31–33, 36
photography and architecture 8, 33, 132, 142–144
photography and social media 111, 139–140, 144
photography, pinhole 15, 34–37
Piano, R. 1, 45–46, 134
Pink, S. 18, 30, 37

resonance 3, 25, 41, 59, 67, 78, 93–94, 102, 124, 146, 153–154
Rose, G. 33, 137

Sennet, R. 7, 24, 51, 122, 148
sketching 8, 17, 31
SLA 11, 121, 123, 125
Sloterdijk, P. 7–8, 15, 26–29, 107, 117, 121, 151–154
Snøhetta 3, 7, 11, 42, 44, 48–51, 71–74, 89–91, 107–109, 131
social media 7–8, 15, 22–23, 68, 126–127, 137–140, 144, 146, 153
social monument 3–5, 70, 81, 100, 146, 148–149, 154–155
Sohlberg, H. 76
Solnit, R. 86, 93
spheres 9, 14, 17, 19, 26–27, 29, 107, 118, 123–124
stains 66–67, 70–71, 154
Statsbygg 42–43, 46–49
stultification 4, 129, 133
synaesthetic 14, 62, 69, 76–80
synchronic space 7
Syndey Opera House 85, 103

Thorsen, K. 69, 71, 77, 86, 89
throwntogetherness 9
time-lapse 8, 33, 83–84, 132
Till, J. 39–41
tourists 7, 44, 68, 99–103, 110, 140

Vergunst, J. L. 32

walking 65, 70, 73, 85–86, 91, 93–97, 101–102
weather 25, 76–78, 81, 89, 94, 96, 99, 103, 107–108, 110, 113–114, 118, 132–135, 149, 154
whiteness 56, 61–62, 67–71, 74–75, 77–81, 140, 154
Wigley, M. 67–68, 71, 76–78
withdrawal, from atmosphere 18, 32, 36–37, 63, 123, 154

Zumthor, P. 3, 24, 152

For Product Safety Concerns and Information please contact our EU representative GPSR@taylorandfrancis.com
Taylor & Francis Verlag GmbH, Kaufingerstraße 24, 80331 München, Germany

www.ingramcontent.com/pod-product-compliance
Ingram Content Group UK Ltd.
Pitfield, Milton Keynes, MK11 3LW, UK
UKHW020002200926
475702UK00015B/2434

* 9 7 8 1 0 3 2 6 6 8 7 8 9 *